Internet of Things with 8051 and ESP8266

Internet of Things with 8051 and ESP8266

Anita Gehlot, Rajesh Singh, Praveen Kumar Malik, Lovi Raj Gupta, Bhupendra Singh

CRC Press
Taylor & Francis Group
Boca Raton London New York

CRC Press is an imprint of the
Taylor & Francis Group, an **informa** business

First edition published 2021 by
CRC Press
6000 Broken Sound Parkway NW, Suite 300, Boca Raton, FL 33487-2742

and by
CRC Press
2 Park Square, Milton Park, Abingdon, Oxon, OX14 4RN

© 2021 Taylor & Francis Group, LLC

CRC Press is an imprint of Taylor & Francis Group, LLC

ISBN: 978-0-367-53478-3 (hbk)
ISBN: 978-1-003-08213-2 (ebk)

Typeset in Times LT Std
by KnowledgeWorks Global Ltd.

Contents

SECTION B   Interfacing of 8051 Microcontroller and NuttyFi/ESP8266 with I/O devices

SECTION C   *Interfacing of 8051 Microcontroller and NuttyFi/ESP8266 with Special Devices*

SECTION D Case Study Based on Data Logger to Cloud Server

Preface

This book will provide a better understanding of 8051 and its interfacing for Internet of Things (IoT) applications. IoT or Internet-enabled communications were simply explained by Jacob Morgan who stated that broadband Internet is widely available and more devices come with Wi-Fi capabilities and sensors with them.

The book includes the basics of IoT, interfacing of 8051, and NuttyFi with I/O devices in different communication modes. The book comprises of four sections and total eighteen chapters.

Authors are thankful to the publisher for the support and encouragement to write this book.

Dr. Anita Gehlot, Lovely Professional University, India
Dr. Rajesh Singh, Lovely Professional University, India
Dr. Praveen Kumar Malik, Lovely Professional University, India
Dr. Lovi Raj Gupta, Lovely Professional University, India
Bhupendra Singh, Schematics Microelectronics, India

Authors' Biographies

Dr. Anita Gehlot is currently associated with Lovely Professional University as Associate Professor with more than 12 years of experience in academics. Her area of expertise includes embedded systems, wireless sensor networks, and IoT. She has been honored as keynote speakers and session chair to national/international conferences, faculty development programs, and workshops. She has 132 patents in her account. She has published more than 70 research papers in referred journals/conferences and 24 books in the area of Embedded Systems and IoT with reputed publishers like CRC/ Taylor & Francis, Narosa, GBS, IRP, NIPA, River Publishers, Bentham Science, and RI Publication. She is editor to a special issue published by AISC book series, Springer, in 2018, and IGI Global in 2019. She has been awarded with "certificate of appreciation" from University of Petroleum and Energy Studies for exemplary work. Under her mentorship, students' team got "InSc Award 2019" under students projects program. She has been awarded with "Gandhian Young Technological Innovation (GYTI) Award", as Mentor to "On Board Diagnostic Data Analysis System-OBDAS", Appreciated under "Cutting Edge Innovation" during Festival of Innovation and Entrepreneurship at Rashtrapati Bhavan, India, in 2018.

Dr. Rajesh Singh is currently associated with Lovely Professional University as Professor with more than 16 years of experience in academics. He has been awarded as gold medalist in M.Tech from RGPV, Bhopal (MP), India, and his B.E. (Hons) from Dr. B.R. Ambedkar University, Agra (UP), India. His area of expertise includes embedded systems, robotics, wireless sensor networks, and IoT. He has been honored as keynote speakers and session chair to national/international conferences, faculty development programs, and workshops. He has 152 patents in his account. He has published more than 100 research papers in referred journals/conferences, and 24 books in the area of Embedded Systems and IoT with reputed publishers like CRC/Taylor & Francis, Narosa, GBS, IRP, NIPA, River Publishers, Bentham Science, and RI Publication. He is editor to a special issue published by AISC book series, Springer, in 2017 & 2018, and IGI Global in 2019. Under his mentorship, students have participated in national/international competitions including "Innovative Design Challenge Competition" by Texas and DST and Laureate award of excellence in robotics engineering, Madrid, Spain, in 2014 & 2015. His team has been the winner of "Smart India Hackathon-2019" hardware version conducted by MHRD, Government of India, for the problem statement of Mahindra & Mahindra. Under his mentorship, students' team got "InSc Award 2019" under students' projects program. He has been awarded with "Gandhian Young Technological Innovation (GYTI) Award", as Mentor to "On Board Diagnostic Data Analysis System-OBDAS", Appreciated under "Cutting Edge Innovation" during Festival of Innovation and Entrepreneurship at Rashtrapati Bhavan, India, in 2018. He has been honored with "Certificate of Excellence" from 3rd faculty branding awards-15, Organized by EET CRS research wing for excellence in professional education and industry, for the category "Award for Excellence in Research", 2015, and young investigator award at the International Conference on Science and Information in 2012.

Dr. Praveen Kumar Malik is working as Professor in Deptt. of Electronics and Communication, Lovely Professional University, Punjab, India. He has more than 15 papers in different national/international journals. His major area of interest includes Antenna Design, Wireless Communication, and Embedded Systems.

Dr. Lovi Raj Gupta is the Executive Dean, Faculty of Technology & Sciences, Lovely Professional University, Punjab, India. He is a leading light in the field of Technical and Higher Education in the country. His research focused approach and an insightful innovative intervention of technology in education has won him much accolades and laurels. He holds a PhD in Bioinformatics. He did his M.Tech in Computer Aided Design & Interactive Graphics from IIT, Kanpur (UP), and B.E. (Hons) from MITS, Gwalior (MP). Having flair for endless learning, has done more than 15 certifications and specializations online on IoT, Augmented Reality and Gamification, from University of California at Irvine, Yonsei University, South Korea, and Wharton School, University of Pennsylvania. His research interests are in the areas of Robotics, Mechatronics, Bioinformatics, IoT, and Gamification.

In 2001, he was appointed as Assistant Controller (Technology), Ministry of IT, Govt. of India, by the Honorable President of India in the Office of the Controller of Certifying Authorities (CCA). In 2013, he was accorded the role in the National Advisory Board for What Can I Give Mission— Kalam Foundation of Dr. APJ Abdul Kalam. In 2011, he received the MIT Technology Review Grand Challenge Award followed by the coveted Infosys InfyMakers Award in the year 2016.

Bhupendra Singh is the Managing Director of Schematics Microelectronics and provides Product design and R&D support to industries and universities. He has completed BCA, PGDCA, M.Sc. (CS), M.Tech, and has more than 11 years of experience in the field of Computer Networking and Embedded Systems. He has published ten books in the area of Embedded Systems and IoT with reputed publishers like CRC/Taylor & Francis, Narosa, GBS, IRP, NIPA, and RI Publication.

Section A

Basics of 8051 Microcontroller and IoT

1 Introduction to the IoT

This chapter discusses the introduction to the Internet of Things (IoT). The chapter also describes the features of IoT and its components.

1.1 INTRODUCTION TO IoT

The meaning of the "Internet" is "The large systems of connected computers around the world that allow people to share information and communicate with each other". "Things" is "used to refer in an approximate way to objects".

The required data can be collected by the IoT enabled device from the existing wide variety of technologies and it can send the data, as and when required, to the identified device. In the present market, smart ACs and heaters act according to the requirements of a Wi-Fi system's user. Old mobile phones, TVs, and house dustbins are becoming smarter. According to one statistical analysis, IoT enabled devices will reach up to 31 billion in number by 2020. IPv6 is a version of Internet protocol, which provides identification for computers on the network and routes the traffic over the Internet. IPv6 has 128-bit Hex numbers.

Features: The basic architecture of the IoT comprises sensors, actuators, and their enabling machine language. Artificial Intelligence, alongside connectivity, and active engagement, can be used by small devices.

Artificial Intelligence: Artificial Intelligence is mathematically developed man-made machine intelligence. It was created in order to use the natural environment so that it can achieve a target. The IoT enabled Artificial Intelligence to develop a smart algorithm in order to collect data and communicate with other connected devices through their networks. For example, in the smart bin system of a production line, if the material moves over, then data will be transferred to an ERP system. This will then be followed by an order received by supplier from ERP, which will refill the intelligent smart bin.

Connectivity: Connectivity is a major issue in most places. Previously, industries had connectivity. At present, XBee, RFID, RX/TX 433MHz, and Wi-Fi are the devices used to provide network connectivity in order to realize the IoT's applications.

Sensors: Devices, in the form of sensors/transducers, are required to detect physical parameters; these then communicate its data to the destination through an embedded system.

Active Engagement: The IoT is an active engagement with technology that makes a paradigm shift over today's passive engagement, which exists in service and product management.

Small Devices: The IoT is a small device that enables and ensures greater precision, scalability, and versatility.

Advantages:

1. In today's scenarios, the IoT is a part of the personal, as well as the business, life of an individual.
2. It improves customer engagement with product service.
3. It is optimized in a way to allow the use of technology.
4. It is easy to collect data collection.

Disadvantages: Although IoT addresses so many meritorious things, but it has some challenges as well:

1. Security: Today everyone is aware of the term "cyber security". This is because most individuals communicate with each other through virtual networks and this puts them at risk of hackers.
2. Complexity: When systems and processes become simpler and more user-friendly, the complexity of developing them increases.
3. Compliance: Any service or technology in business needs to be complying with regulations.

The IoT is the integration of sensation, communication, and analytical capabilities created within conventional technologies. The IoT promises to help the automotive industry by directly managing their existing assets in different places. This allows information from the supply chain and after sales services, as well as dealers and customers, to be collated in order to help them to understand and access the data/information as and when required.

1.1.1 IoT in Society

Society is becoming smarter by using the IoT ecosystem in day-to-day activities; this has enriched the lives of human beings. Its application began with the smart dustbin and has progressed to operating the garage door. In the near future, the IoT could become a multi-trillion dollar industry. CISCO revealed that the use of IoT enabled devices may increase to 50 billion devices used by 7.6 billion people.

1.2 IoT COMPONENTS

The system used by the IoT comprises of many functional blocks to facilitate various utilities such as, sensing, identification, actuation, communication, and management.

Figure 1.1 shows the various components of IoT devices such as the connectivity, processor, audio/video interfaces, input-output interfaces, storage interface, memory interface, and graphics.

1.2.1 Sensor

The sensor is a device that collects information from the environment and converts it into the electrical signals. The type of a sensor depends on the application as well as its availability in the market. These can be classified in two types: analog and digital. The digital sensors are also available

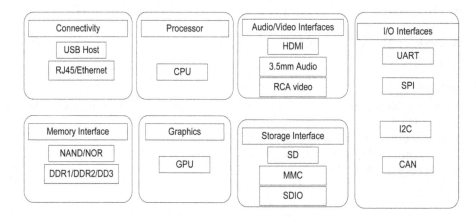

FIGURE 1.1   IoT device components.

as inter IC communication (I²C), two-wire interface (TWI), serial peripheral interface (SPI), three-wire interface, or one wire interface.

1.2.2   PUSH BUTTON

A push button is used to detect the physical contact; for example, bumpers on a robot can be equipped with the push buttons. When the button is closed (pressed), it makes a connection between two terminals and can be used for sending a signal to do a specific task.

1.2.3   PRESSURE SENSOR

A pressure sensor produces an output proportional to the force that is being applied to it.

1.2.4   ULTRASONIC SENSOR

An ultrasonic sensor is an acoustics sensor that is capable of measuring the time between, when a signal is sent and when its echo is received. It can be used to measure the distance between an object and an obstacle. The ultrasonic sensor is an analog sensor with a range of 10–400 cm.

1.2.5   INFRARED SENSOR

An infrared (IR) sensor is capable of detecting an obstacle in the line follower application. The IR sensor can be used for the collision detection. The module comprises of an IR transmitter and IR receiver pair, a comparator IC-358, and a potentiometer.

1.2.6   PROXIMITY AND TOUCH SENSOR

This is a digital sensor. The principle of this sensor is to detect a change in capacitance, when a finger is near the pad. The sensor has two types of output. The first output is called a proximity output, which is activated when a finger is around 2cm above the pad. The other output is a touch output, which is activated when a finger is either on the pad or around 1mm above.

1.2.7   LASER SENSOR

A laser sensor can be used in long distance application, which needs a high degree of accuracy.

1.2.8   BUMP SENSOR

This is a digital sensor and can be used for the obstacle detection. It has an onboard performance limit switch and an anti-debouncing circuit.

1.2.9   METAL SENSOR

This is a digital sensor that gives an active low output when it detects metal. It detects metal objects from up to 7cm away.

1.2.10   THE FIRE SENSOR

As the name suggests, this is used for detecting fire. The module detects the fire up to a range of 1–2 meters.

1.2.11 The Passive Infrared (PIR) Sensor

This is a digital sensor that is designed to detect the presence of people. It can be operated on +5V to +9V DC.

1.2.12 Potentiometer

A potentiometer is a simple knob which provides a variable resistance.

1.2.13 Alcohol Sensor

This is used to detect an alcohol on breath or in air. It is highly sensitive with a fast response time. Its operating temperature range is -10–70°C.

1.2.14 Temperature Sensor

An LM35 is a temperature sensor, whose output voltage is linearly proportional to Celsius (Centigrade). An LM35 is designed to operate for a temperature range of -55–150°C.

1.2.15 Gas Sensor

This is used to detect gas for liquefied petroleum gas (LPG), isobutane, propane, and liquefied natural gas (LNG) combustible gases. The detecting range is 100–100,00PPM.

1.2.16 GPS

The Global Positioning System (GPS) is a global navigation satellite system. It uses a constellation of between 24 and 32 "Medium Earth Orbit" satellites in order to determine location, speed, direction, and time. GPS can be used for navigation by measuring latitude, longitude, and the altitude of the system.

1.2.17 Gyroscope

A gyroscope measures the rate of an angular acceleration and provides a corresponding signal (analog voltage, serial communication, I²C, and so on).

1.2.18 Thermal Camera

This is used to create a 2-D image of objects. The temperature of an object can also be determined by using this device.

1.2.19 Humidity Sensor

The humidity sensor detects the percentage of water in the air and can be used with the temperature sensor.

1.2.20 Magnetometer

The magnetic sensors or magnetometers can be used to detect the magnets and magnetic fields. This can be used to find out the position of magnets.

1.2.21 COMPASS SENSOR

This is used to determine the earth's magnetic field in order to find out its orientation with respect to the magnetic poles.

1.2.22 ACCELEROMETER

An accelerometer measures the linear acceleration. It can measure the static (earth gravity) or dynamic acceleration in three axes.

1.2.23 LOAD CELL

This is a transducer that converts a force into an electrical signal. The output electrical signal is a few millivolts and requires amplification.

1.2.24 FINGERPRINT SENSOR

This is used to read fingerprints and store the data in its flash memory. The sensor can perform three functions: Add (Enroll), Empty Database, and Search. It operates on +5V DC. For every fingerprint it generates a 32-bit data string.

1.2.25 INERTIAL MEASUREMENT UNIT

An inertial measurement unit is a combination of a multi-axis accelerometer and a multi-axis gyroscope and, sometimes a multi-axis magnetometer, to provide more accuracy.

1.2.26 CURRENT AND VOLTAGE SENSOR

A current and voltage sensor measures the current and/or voltage of an electric circuit. This can be useful for gauging how much longer a robot can be operated (i.e., it can measure the voltage from the battery).

1.2.27 VIBRATION SENSOR

A vibration sensor can be used to detect the vibration of an object by using piezoelectric or other technologies.

1.2.28 RFID

A RFID module can be interfaced with a microcontroller in UART mode or with an RS232 converter to a PC. A RFID reader module works with 125 KHz RFID tags. It operates on +5V DC power supply and it has a range of 10cm.

1.3 ACTUATORS

The actuator can be defined as a device, which converts electrical energy into motion. The different variety of actuators can produce either rotational or linear motions. A DC motor is an example of an actuator. The selection of the right actuator for a robot requires an understanding of which actuator is available and suitable; this can be worked out using a bit of math and physics.

1.3.1 ROTATIONAL ACTUATORS

The rotational actuator transforms the electrical energy into a rotating motion. The properties of a rotational actuator are—the torque (the force an actuator produces at a given distance [unit is N•m]) and its rotational speed (unit in rpm).

1.3.2 AC MOTOR

An AC motor is not used in the robotics. AC motors are useful for industrial purposes with a stationary arrangement, where very high torque is required; the motors are connected to the mains/wall outlet.

1.3.3 DC MOTOR

The DC motors are available in various shapes and sizes. The most common shape is cylindrical. A DC motor can operate in both clockwise and anti-clockwise direction by changing the polarity of the battery. These motors can be purchased from the market; they differ in their operating voltage (in volt) and speed (rpm).

1.3.4 GEARED DC MOTOR

A geared DC motor is a combination of a gearbox and the DC motor, which can decrease the speed and increase the torque of a motor. The types of gears are, the spur gear (most commonly used), the planetary gear(more complex, but it allows the higher gear down to be operated in a more confined space and with higher efficiency), and the worm gear (this allows a very high gear ratio in a single stage, and also prevents the output shaft from moving, if the motor is not powered).

1.3.5 SERVO MOTOR

The servo motor is a type of the actuator, which rotates to an angular position and can be used in the remote-controlled vehicles for steering or controlling the flight. It has three wires, GND, Vcc, and a control input pulse.

1.3.6 INDUSTRIAL SERVO MOTOR

The control of an industrial servo motor is different from a hobby servo motor. It is used in large machines. An industrial servo motor is made up of an AC motor and, sometimes, a three-phase motor. It is designed by placing a gear down and an encoder, which provides the feedback about its angular position and speed. These motors are rarely used for the mobile robots due to their heavy weight, size, cost, and the complexity. These are more useful in the stationary robots.

1.3.7 STEPPER MOTOR

A stepper motor rotates in the steps with a minimum step angle of 1.8 degree. The number of degrees the shaft rotates with each step (step size) varies based on several factors. There are two types of stepper motor: unipolar and bipolar.

1.3.8 LINEAR ACTUATOR

A linear actuator is used where a linear motion (motion along one straight line) is required. It has three main mechanical characteristics:

1. The minimum and maximum distance that the rod can move, in mm or inches
2. The force in kg or lbs
3. The speed in m/s or inches

1.3.9   DC Linear Actuator

A DC linear actuator is made of a DC motor connected to a lead screw. The forced motion of lead screw either toward or away from the motor, converts the rotating motion into a linear motion. The DC linear actuators provide linear position feedback by connecting with a linear potentiometer. These are available in a wide variety of sizes, strokes, and forces.

1.3.10   Solenoid

A solenoid is designed with a coil wound around a core. When the coil is energized, the core is pushed away from the magnetic field and this result in a single directional motion. The multiple coils, or some other mechanical arrangements, are required to provide a motion in two directions. This type of the actuator can be used in valves or latching systems where no position feedback is required.

1.3.11   Muscle Wire

The muscle wire is a special type of wire that can contract when an electric current is moving through it; once the current is gone, it returns to its original length.

1.3.12   Pneumatic and Hydraulic Actuators

In order to produce a linear motion, the pneumatic and hydraulic actuators use air or a liquid such as water or oil. These types of actuators can have a very long stroke, a high force, and a high speed.

1.4   CLOUD COMPUTING MODEL

Connection and accessibility are crucial in the modern business world, and cloud computing allows the business to work wherever you like, or whenever you like. However, cloud computing is not quite as simple as it may seem. Inside of the world of cloud computing, there are three major service models. By comparing the three different models, you will be better able to determine which cloud computing service model is right for your business.

Cloud Computing Service Models

The three types of cloud models are Software-as-a-Service, Platform-as-a-Service, and Infrastructure-as-a-Service. Their basic functions can be summarized in the phrases "Host", "Build", and "Consume". Each offers a different level of flexibility and control over the product that your business is "buying". Each also varies in its relationship to your existing IT infrastructure. Because of the wide variances between the three, it is important to determine which model will suit your business needs the best.

1.4.1   Software as a Service (SaaS)

The SaaS model allows your business to quickly access cloud-based web applications without committing to installing new infrastructure. The applications run on the vendor's cloud, which they, of course, control and maintain. The applications are available for use with a paid licensed subscription, or for free with limited access. SaaS does not require any installations or downloads in your existing infrastructure, which in turn eliminates the need to install, maintain, and update applications on each of your computers.

Advantages of SaaS

Affordable

On-premise hardware is not required for this model, which keeps the costs associated low. Small-scale businesses might find this cloud platform particularly appealing.

Accessible Everywhere

Cloud-based applications are accessible everywhere that there is internet access. As such, companies that require frequent collaboration find SaaS platforms useful as their employees can easily access the programs that they need.

Ready-to-Use

With SaaS, the programs you need are already fully developed and ready to use. The set-up time for SaaS programs is greatly decreased from the other two types of cloud-based platforms.

Disadvantages of SaaS

Lack of Control

With SaaS, the vendor has control over the programs that your company is using. If you do not feel comfortable releasing the control of your critical business applications to another party, perhaps SaaS is not the best option for your business.

Slower Speeds

Relying upon internet access to function, SaaS applications tend to be slower than client/server applications. However, these programs are still typically quick, though not instantaneous.

Variable Functions and Features

In many cases, SaaS cloud-based applications have less functionality and features than their client/server counterparts. This disadvantage, however, may be void if your business only needs the features offered in the SaaS version to function.

1.4.2 PLATFORM AS A SERVICE (PAAS)

With this model, a third-party vendor provides your business with a platform upon which your business can develop and run applications.

Because the vendor is hosting the cloud infrastructure, which supports the platform, PaaS eliminates your need to install in-house hardware or software. Your business would not manage or control the underlying cloud infrastructure, but you would maintain control over the deployed applications (unlike with SaaS).

Advantages of PaaS

Rapid Time-to-Market

PaaS simplifies application management by eliminating the need to maintain and control the underlying infrastructure. As a result, applications can be developed and deployed faster.

Cost-Effective Development

A cloud-based platform provides your business with a base upon which to build your applications, as opposed to building from nothing, thus dramatically reducing the costs associated.

Scalability

Cloud-based platforms offer reusable code, which, of course, makes it easier to develop and deploy applications, but also offers increased scalability.

Disadvantages of PaaS

Vendor Lock-In

It is difficult to migrate many of the services provided by one PaaS product to a competing product, thus making it hard to switch PaaS vendors. Downtime and additional expenses are likely to occur when switching from one PaaS provider to another.

Security and Compliance

In the PaaS model, the vendor will store most, or even all, of the application's data. As such, it is imperative to assess the security measures of the provider. This, though, often proves difficult as the vendor may be storing their databases via a third party, thus leaving you uninformed of the safety of your data.

Lack of Compatibility

It is possible that your current infrastructure may not be compatible with a cloud platform. If some elements cannot be cloud-enabled, you may have to switch from your current apps and programs to cloud-compatible counterparts in order to fully integrate. Alternately, you may need to leave these elements out of the cloud, and within your current infrastructure.

1.4.3 Infrastructure as a Service (IaaS)

IaaS, as the most flexible of the cloud models, allows your business to have complete, scalable control over the management and customization of your infrastructure.

In the IaaS model, the cloud provider hosts your infrastructure components that would traditionally be present in an onsite data center (such as servers, storage, and networking hardware). Your business, however, would maintain control over operating systems, storage, deployed applications, and possibly limited control of select networking components (e.g., host firewalls).

Advantages of IaaS

Eliminates Capital Expenses

Employing a cloud-based infrastructure eliminates the capital expense of deploying in-house hardware and software. Additionally, IaaS typically is offered as a pay-as-you-go model, with charges based either in time, or in the amount of virtual machine space that was used.

Supports Flexibility

IaaS is useful in supporting workloads that are temporary, may change unexpectedly, or are experimental. Like all workloads, these loads need infrastructure to support them; however, it is expensive to commit to additional permanent in-house infrastructure for a temporary need. Cloud-based infrastructure answers the need for flexibility.

Simple Deployment

It is much easier for your cloud provider to deploy your servers, processing, storage, and networking in the IaaS model, than it is for you to deploy these elements in-house, with no previous base to build. As a result, your uptime will increase as your systems will be available for use more rapidly.

Disadvantages of IaaS

Insight

Because your entire infrastructure is maintained and controlled by your IaaS provider, it is rare that you will be provided with the details of its configuration and performance. In turn, this can make systems management and monitoring more difficult for your company.

Variability of Resilience

The availability and performance of the workload is highly dependent upon the provider. If the IaaS providers experience internal or external downtime, your workloads will also be affected.

Costly

IaaS models are typically much more costly than PaaS and SaaS models because they offer much more support to your business than the other two cloud models. However, they can still be cost-effective based on their utility to your business.

1.5   ZIGBEE

Zigbee is an IEEE 802.15.4-based specification for a suite of high-level communication protocols used to create personal area networks with small, low-power digital radios, such as for home automation, medical device data collection, and other low-power low-bandwidth needs, designed for small scale projects, which need wireless connection. Hence, Zigbee is a low-power, low data rate, and close proximity (i.e., personal area) wireless ad hoc network.

The technology defined by the Zigbee specification is intended to be simpler and less expensive than other wireless personal area networks (WPANs), such as Bluetooth or more general wireless networking such as Wi-Fi. Applications include wireless light switches, home energy monitors, traffic management systems, and other consumer and industrial equipment that require short-range low-rate wireless data transfer.

Its low-power consumption limits transmission distances to 10–100 meters line-of-sight, depending on power output and environmental characteristics. Zigbee devices can transmit data over long distances by passing data through a mesh network of intermediate devices to reach more distant ones. Zigbee is typically used in low-data rate applications that require long battery life and secure networking (Zigbee networks are secured by 128-bit symmetric encryption keys.) Zigbee has a defined rate of 250 kbit/s, best suited for intermittent data transmissions from a sensor or input device.

Zigbee is a low-cost, low-power, wireless mesh network standard targeted at battery-powered devices in wireless control and monitoring applications. Zigbee delivers low-latency communication. Zigbee chips are typically integrated with radios and with microcontrollers. Zigbee operates in the industrial, scientific, and medical (ISM) radio bands: 2.4 GHz in most jurisdictions worldwide; though some devices also use 784 MHz in China, 868 MHz in Europe, and 915 MHz in the US and Australia; however, even those regions and countries still use 2.4 GHz for most commercial Zigbee devices for home use. Data rates vary from 20 kbit/s (868 MHz band) to 250 kbit/s (2.4 GHz band).

Zigbee builds on the physical layer and media access control defined in IEEE standard 802.15.4 for low-rate WPANs. The specification includes four additional key components: network layer, application layer, Zigbee Device Objects (ZDOs), and manufacturer-defined application objects. ZDOs are responsible for some tasks, including keeping track of device roles, managing requests to join a network, as well as device discovery and security.

The Zigbee network layer natively supports both star and tree networks, and generic mesh networking. Every network must have one coordinator device. Within star networks, the coordinator must be the central node. Both trees and meshes allow the use of Zigbee routers to extend communication at the network level. Another defining feature of Zigbee is facilities for carrying out secure communications, protecting establishment and transport of cryptographic keys, ciphering frames, and controlling device. It builds on the basic security framework defined in IEEE 802.15.4.

1.6   BLE

Bluetooth Low Energy (Bluetooth LE, colloquially BLE, formerly marketed as Bluetooth Smart[1]) is a WPAN technology designed and marketed by the Bluetooth Special Interest Group (Bluetooth SIG) aimed at novel applications in the healthcare, fitness, beacons, security, and home entertainment industries. It is independent of Bluetooth BR/EDR and has no compatibility, but BR/EDR and LE can coexist. The original specification was developed by Nokia in 2006 under the name Wibree, which was integrated into Bluetooth 4.0 in December 2009 as BLE.

Compared to Classic Bluetooth, BLE is intended to provide considerably reduced power consumption and cost while maintaining a similar communication range. Mobile operating systems including iOS, Android, Windows Phone, and BlackBerry, as well as macOS, Linux, Windows 8, and Windows 10, natively support BLE.

In 2011, the Bluetooth SIG announced the Bluetooth Smart logo so as to clarify compatibility between the new low energy devices and other Bluetooth devices.

1.7   6LoWPAN

6LoWPAN is an acronym of IPv6 over Low-Power WPANs. 6LoWPAN is the name of a concluded working group in the Internet area of the IETF.

The 6LoWPAN concept originated from the idea that "the Internet Protocol could and should be applied even to the smallest devices", and that low-power devices with limited processing capabilities should be able to participate in the IoT.

The 6LoWPAN group has defined encapsulation and header compression mechanisms that allow IPv6 packets to be sent and received over IEEE 802.15.4 based networks. IPv4 and IPv6 are the work horses for data delivery for local-area networks, metropolitan area networks, and wide-area networks such as the Internet. Likewise, IEEE 802.15.4 devices provide sensing communication-ability in the wireless domain. The inherent natures of the two networks though, are different.

The base specification developed by the 6LoWPAN IETF group is RFC 4944 (updated by RFC 6282 with header compression and by RFC 6775 with neighbor discovery optimizations). The problem statement document is RFC 4919. IPv6 over BLE is defined in RFC 7668.

As with all link-layer mappings of IP, RFC4944 provides a number of functions. Beyond the usual differences between L2 and L3 networks, mapping from the IPv6 network to the IEEE 802.15.4 network poses additional design challenges (see RFC 4919 for an overview).

1.7.1   FUNCTIONS

Adapting the Packet Sizes of the Two Networks

IPv6 requires the maximum transmission unit (MTU) to be at least 1280 octets. In contrast, IEEE 802.15.4's standard packet size is 127 octets. A maximum frame overhead of 25 octets spares 102 octets at the media access control layer. An optional but highly recommended security feature at the link layer poses an additional overhead. For example, 21 octets are consumed for AES-CCM-128 leaving only 81 octets for upper layers.

Address Resolution

IPv6 nodes are assigned 128 bit IP addresses in a hierarchical manner, through an arbitrary length network prefix. IEEE 802.15.4 devices may use either of IEEE 64 bit extended addresses or, after an association event, 16 bit addresses that are unique within a PAN. There is also a PAN-ID for a group of physically collocated IEEE 802.15.4 devices.

Differing Device Designs

IEEE 802.15.4 devices are intentionally constrained in form factor to reduce costs (allowing for large-scale network of many devices), reduce power consumption (allowing battery powered devices), and allow flexibility of installation (e.g., small devices for body-worn networks). On the other hand, wired nodes in the IP domain are not constrained in this way; they can be larger and make use of mains power supplies.

Differing Focus on Parameter Optimization

IPv6 nodes are geared toward attaining high speeds. Algorithms and protocols implemented at the higher layers such as TCP kernel of the TCP/IP are optimized to handle typical network problems such as congestion. In IEEE 802.15.4-compliant devices, energy conservation and code-size optimization remain at the top of the agenda.

Adaptation Layer for Interoperability and Packet Formats

An adaptation mechanism to allow interoperability between IPv6 domain and the IEEE 802.15.4 can best be viewed as a layer problem. Identifying the functionality of this layer and defining newer packet formats, if needed, is an enticing research area. RFC 4944 proposes an adaptation layer to allow the transmission of IPv6 datagrams over IEEE 802.15.4 networks.

Addressing Management Mechanisms

The management of addresses for devices that communicate across the two dissimilar domains of IPv6 and IEEE 802.15.4 is cumbersome, if not exhaustingly complex.

Routing Considerations and Protocols for Mesh Topologies in 6LoWPAN

Routing per se is a two-phased problem that is being considered for low-power IP networking:
   Mesh routing in the personal area network (PAN) space.

The Routability of Packets between the IPv6 Domain and the PAN Domain

Several routing protocols have been proposed by the 6LoWPAN community such as LOAD, DYMO-LOW, and HI-LOW. However, only two routing protocols are currently legitimate for large-scale deployments: LOADng standardized by the ITU under the recommendation ITU-T G.9903 and RPL standardized by the IETF ROLL working group.

Device and Service Discovery

Since IP-enabled devices may require the formation of ad hoc networks, the current state of neighboring devices and the services hosted by such devices will need to be known. IPv6 neighbor discovery extension is an internet draft proposed as a contribution in this area.

Security

IEEE 802.15.4 nodes can operate in either secure mode or non-secure mode. Two security modes are defined in the specification in order to achieve different security objectives: Access Control List (ACL) and Secure mode.

2 Meet 8051 and Keil Compiler—A Software Development Environment

This chapter discusses the basics of microcontroller 8051, with the help of block diagram and pin description of 8051. It also introduces Keil compiler to program the 8051.

2.1 INTRODUCTION TO 8051

8051 microcontroller was designed by Intel in 1981. It is an 8-bit microcontroller. It is built with 40 pins, 4kb of ROM storage and 128 bytes of RAM storage, two 16-bit timers. It consists of four parallel 8-bit ports, which are programmable as well as addressable as per the requirement. An on-chip crystal oscillator is integrated in the microcontroller having crystal frequency of 12MHz or 11.0592MHz. In the following diagram, the system bus connects all the support devices to the CPU. The system bus consists of an 8-bit data bus, a 16-bit address bus, and bus control signals. All other devices like program memory, ports, data memory, serial interface, interrupt control, timers, and the CPU are all interfaced together through the system bus. Figure 2.1 shows the block diagram of 8051.

The pin diagram of 8051 microcontroller is shown in Figure 2.2.

2.2 KEIL COMPILER—A SOFTWARE DEVELOPMENT ENVIRONMENT

Keil Software makes C compilers, macro assemblers, real-time kernels, debuggers, simulators, integrated environments, and evaluation boards for the 8051, 251, ARM, and XC16x/C16x/ST10 microcontroller families. Keil doesn't make a compiler for every different microcontroller architecture. It concentrates on just a few chip families, which can support different architecture very well. That's why the most of Keil's customers will agree that they have the best development tools for ARM, XC1 6x/C1 6x/ST1 0, 251, and 8051 microcontroller families. Since they have fewer product lines to support, Keil can dedicate more time to better supporting each tool chain.

2.2.1 uVision3 IDE

uVision3, the new IDE from Keil Software, combines project management, make facilities, source code editing, program debugging, and complete simulation in one powerful environment, pVision3 helps to get programs working faster than ever while providing an easy-to-use development platform. The editor and debugger are integrated into a single application and provide a seamless embedded project development environment.

uVision3 provides unique features like the following.

The Device Database, which automatically sets the assembler, compiler, and linker options for the chip you select. This prevents you from wasting your time configuring the tools and helps you get started writing code faster.

A robust Project Manager who lets you to create several different configurations of your target from a single project file. Only the Keil pVision3 IDE allows you to create an output file for simulating, an output file for debugging with an emulator, and an output file for programming an EPROM—all from the same project file.

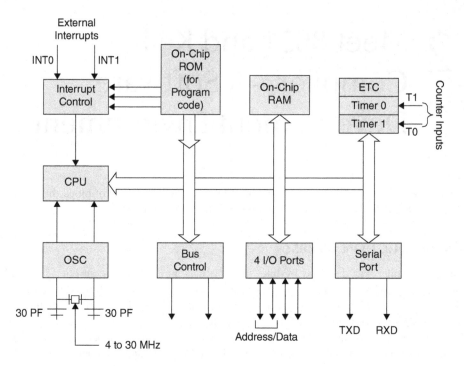

FIGURE 2.1   Block diagram of 8015 microcontroller.

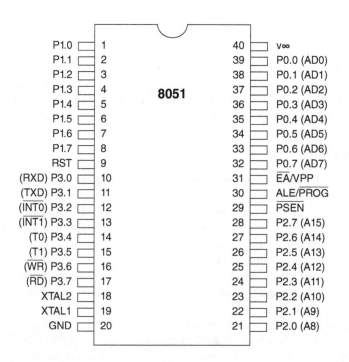

FIGURE 2.2   8051 pin diagram.

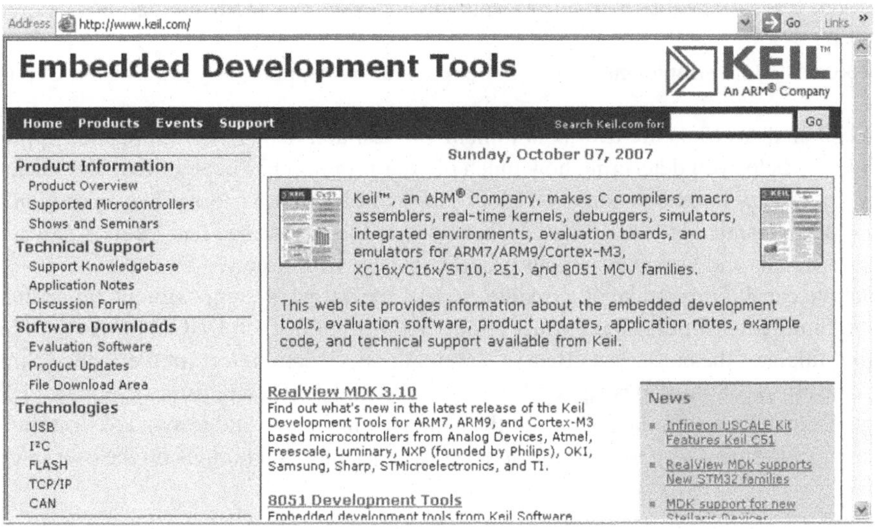

FIGURE 2.3 View of Keil compiler.

It has an integrated make facility with automatic dependency generation. You don't have to figure out which header files and include files are used by which source files. The Keil compilers and assemblers do that automatically.

As your project compiles, errors and warnings appear in an output window. You may make corrections to the files in your project while pVision3 continues to compile in the background. Line numbers associated with each error or warning are automatically resynchronized when you make changes to the source. Figure 2.3 shows the Keil compiler view.

2.2.2 uVision Debugger

The uVision Debugger from Keil supports simulation using only your PC or laptop, and debugging using your target system and a debugger interface. uVision includes traditional features like simple and complex breakpoints, watch windows, and execution control as well as sophisticated features like trace capture, execution profiler, code coverage, and logic analyzer. The uVision Debugger provides a number of ways to display variables and program objects. Source Code Windows display your high-level language and assembly program source code. The Disassembly Window shows mixed high-level language and assembly code. The Registers Tab of the Project Workspace shows system registers. The Symbol Window hierarchy displays program symbols in your application. The Output Window displays the output of various debugger commands. The Memory Window displays up to four regions of code or data memory. The Watch Window displays local variables, user-defined watch expression lists, and the call stack.

2.2.3 Execution

uVision offers many ways by which you can control and manipulate program execution.

Reset—It is possible to debug reset conditions using the uVision simulator.
Run/Stop—Buttons and Commands make starting and stopping program execution easy.
Single-Stepping—uVision supports various methods of single-stepping through your target program.

Execution Trace—Execution trace information for each executed instruction is stored by uVision.

Breakpoints—Both simple and complex breakpoints are supported by the uVision Debugger.

Advanced analysis tools are available to help you test and debug your embedded applications. Code Coverage helps you determine how much of your program has been tested. The Performance Analyzer shows how functions and code blocks in your program perform. The Execution Profiler shows execution counts and time for each line of code or instruction. The Logic Analyzer shows how various signals and variables in your program change over time.

Simulation capabilities make it possible to test the target system without target hardware. Instruction Simulation simulates the exact effects and timing of each IVICU instruction. Interrupt Simulation simulates the cause and effect of a system or peripheral interrupt. Peripheral Simulation simulates the effects of on-chip peripherals including special function registers. Debugger Functions allow you to expand the command scope of the debugger and create and respond to stimuli. Toolbox Buttons are a convenient way for you to connect debugger function buttons on the user-interface.

3 Introduction to NuttyFi Board and Its Programming

Nuttify is an Internet of Things platform with on board ESP122 8266 series. This chapter discusses the NuttyFi board, its specification, and the programming steps for the same.

3.1 INTRODUCTION TO NUTTYFI BOARD

NUTTYFI is an Internet of Things (IoT) hardware platform based on ESP12e 8266 series that enables user to build IoT products, Research Analysis Systems, Automation, and Projects.

Using NUTTIFY, users can monitor, manage, control, and search devices from any part of the world. It can also interface with any IoT web server, icloud, Local or IoT mobile platform or application easily. A number of services like open source web servers and mobile apps are available on the internet that are freeware to use.

In order to make easier for developers, NUTTYFI appears to be a complete and best solution for IoT Products & Projects. NUTTYFI IoT hardware has such ability that the user can upload their code as per their requirements. Using NUTTYFI, you will be capable to control your application, devices, actuators, mechanism as well as collect data from IoT devices securely. Figure 3.1 shows the front view of NuttyFi, while Figure 3.2 shows the back view of NuttyFi with pins.

3.1.1 PIN DESCRIPTION

1. 8 digital pins: From D0 to D7
2. 1 Analog pin-A0
3. Vinput: From 5V to 21V
4. 3.3V output pin
5. UART Pins to Flash program to NUTTYFI Cloud Device using FTDI UART Bridge.

3.1.2 SPECIFICATION OF NUTTYFI

- Voltage: Input from 5V to 21V DC
- Wi-Fi Direct (P2P), soft-AP
- Current consumption: 10uA~170mA
- Flash memory attachable: 16MB max. (512K normal)
- Integrated TCP/IP protocol stack
- Processor: Tensilica L106, 32-bit
- Processor speed: 80~160MHz
- RAM: 32K + 80K
- GPIOs: 17 (multiplexed with other functions)
- Analog to Digital: 1 input with 1024 step resolution
- +19.5dBm output power in 802.11b mode
- 802.11 support: b/g/n
- Maximum concurrent TCP connections: 5.

FIGURE 3.1 Front view of NuttyFi board.

3.2 PROGRAMMING STEPS TO NUTTYFI/ESP8266

3.2.1 INSTALLING NUTTYFI IN TO ARDUINO IDE

The most basic way to use the ESP8266 module is to use serial commands, as the chip is basically a Wi-Fi/Serial transceiver.

Follow the steps:

1. Open the Arduino IDE. Arduino IDE version must be 1.6.4 or greater.
2. Note that your Laptop/Computer must be connected with Internet to install NUTTYFI in Arduino IDE.
3. Click on File and then Click on Preferences, as shown in Figure 3.3.
4. Preference window will appear, copy and paste link to Additional Boards Manager, Figure 3.4. URL is as follows:
   http://arduino.esp8266.com/stable/package_esp8266com_index.json

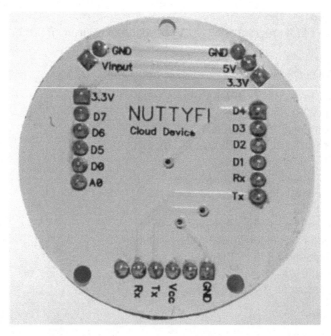

FIGURE 3.2 Back view of NuttyFi with pins.

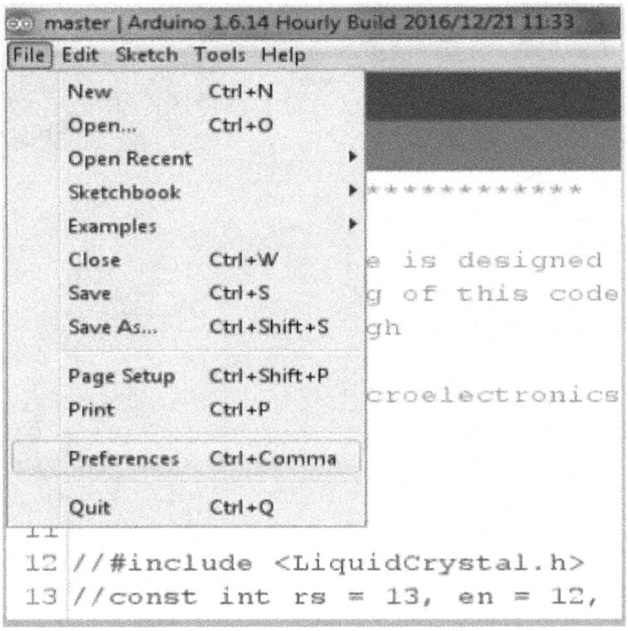

FIGURE 3.3 Select "Preferences".

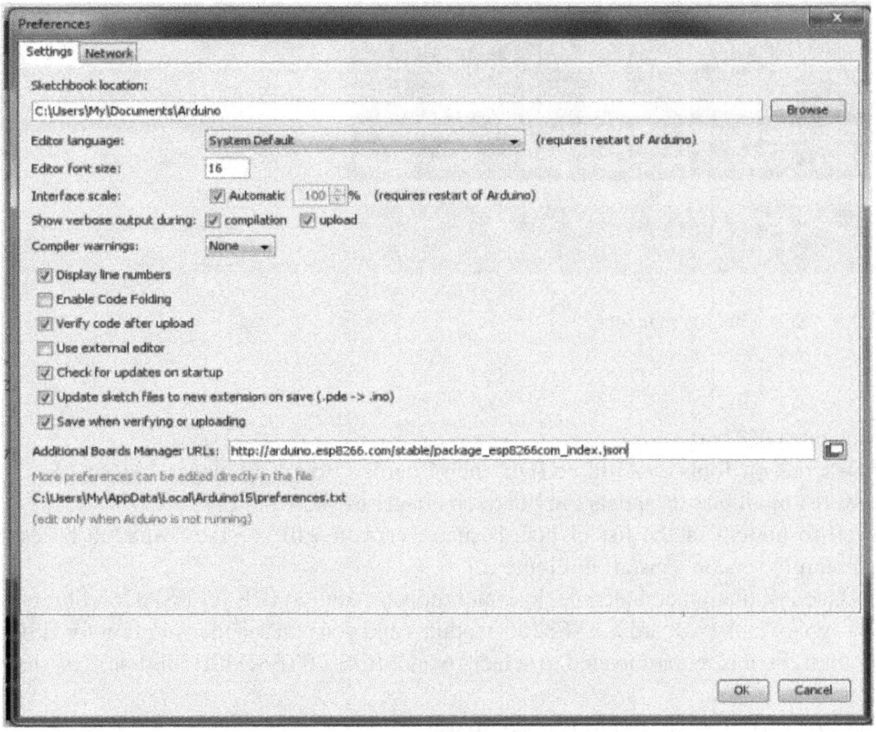

FIGURE 3.4 URL to the additional boards manager.

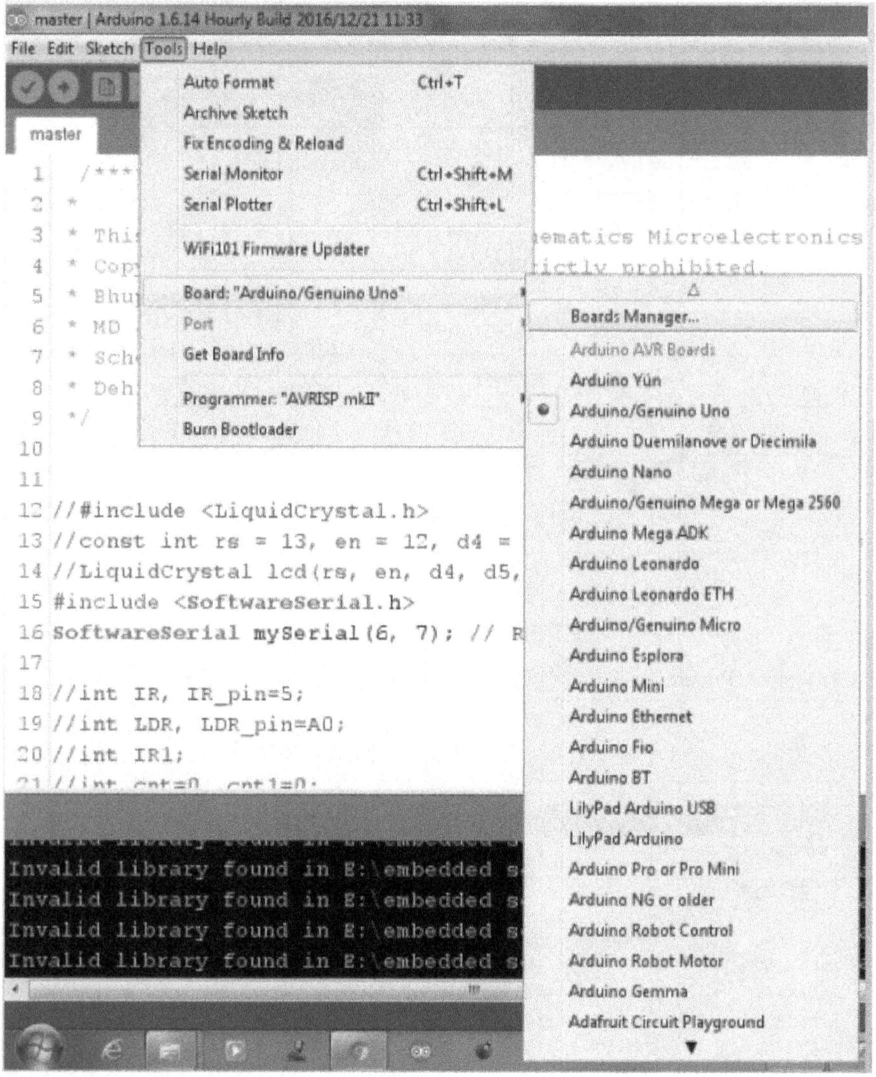

FIGURE 3.5   Select boards manager.

5. Click on "OK" button.
6. Now, Click on Tools on Arduino IDE, then Boards> Boards Manager, Figure 3.5.
7. Now, list of all boards appears in boards manager as shown in Figure 3.6.
8. Scroll to bottom of the list of boards manager, you will see the "esp8266 by esp8266 community version", install it, Figure 3.7.
    Once installation completed, close and reopen Arduino IDE for ESP8266 library.
9. Now, go to Tools > Board > ESP8266 Modules and you can see many option for ESP8266. For NuttyFi, it is recommended to select "NodeMCU 1.0 (ESP-12E Module)", as shown in Figure 3.8.
10. Now NUTTYFI Wi-Fi Board is ready to use.
11. Next, select the port. If you can't recognize your port, go to the Control Panel > System > Device Manager > Port and update your USB driver.

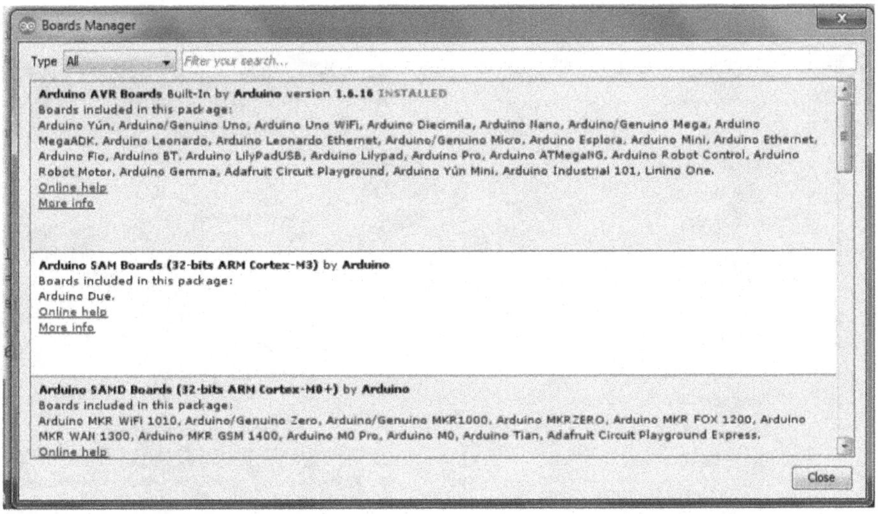

FIGURE 3.6 List of boards.

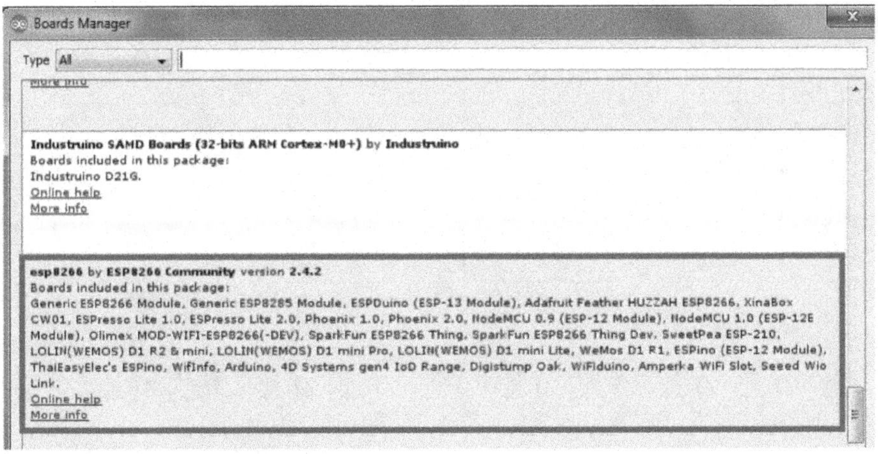

FIGURE 3.7 Installing ESP8266.

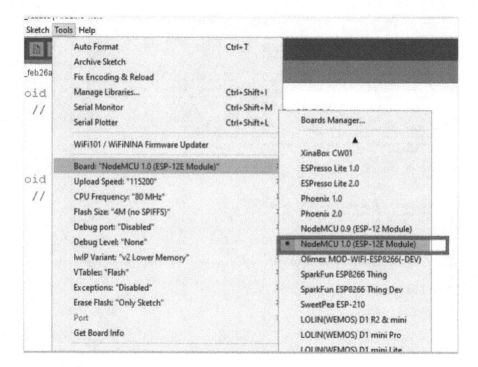

FIGURE 3.8   Select "NodeMCU1.0 (ESP-12E Module)".

4 Introduction to Customized Board with 8051 Microcontroller and NuttyFi/ESP8266

This chapter discusses the development of a customized board with 8051 and NuttyFi for Internet of things (IoT) applications. The brief description of each component is discussed.

4.1 INTRODUCTION TO CUSTOMIZED BOARD WITH 8051 MICROCONTROLLER AND NUTTYFI/ESP8266

The customized board comprises of 89S51 microcontroller, reset circuitry, crystal circuitry with 11.0592MHz frequency crystal, ADC 0804, MUX 74HC4051, and NuttyFi board/ESP8266. Figure 4.1 shows the schematics of customized board.

The description of the components is as follows:

4.1.1 ADC 0804 (ADC OR ANALOG-TO-DIGITAL CONVERSION)

ADC is used to convert analog signals to digital data. ADC 0804 is a chip designed to convert analog signal into 8-bit digital data. This chip is one of the popular series of ADC, Figure 4.2.

This chip is specially designed for getting digital data for processing units from analog sources. It is an 8-bit conversion unit, with a measuring voltage of maximum value 5V. It can sense minimum 4.8mV.

The input analog signal has a limit to its value. This limit is determined by reference value and chip supply voltage. The measuring voltage cannot be greater than reference voltage and chip supply voltage. If the limit is crossed, Vin > Vref, the chip gets faulted permanently.

Pin 9 is Vref/2 that means say we want to measure an analog parameter with a maximum value of 5V. If we need Vref as 5V for that we need to provide a voltage of 2.5V (5V/2) at the Pin 9.

For 2.5V a voltage divider is used, with same value of resistor at both ends they share voltage equally. Each resistor holds a drop of 2.5V with a supply voltage of 5V. The drop from the later resistor is taken as a Vref.

The chip works on RC (Resistor Capacitor) oscillator clock. The important point is the capacitor can be changed to a lower value for higher rate of ADC conversion. However, with speed there will be a decrease in accuracy.

So, if the application requires higher accuracy choose the capacitor with higher value. For higher speed, choose lower value capacitor. On 5V ref., if an analog voltage of 2.3V is given for ADC conversion we will have 2.3*(1024/5) = 471. This will be the digital output of ADC 0804.

4.1.2 MUX 74HC4051

The 74HC4051, 74HCT4051 is a single-pole octal-throw analog switch (SP8T) suitable for use in analog or digital 8:1 multiplexer/de-multiplexer applications. The switch features three digital select

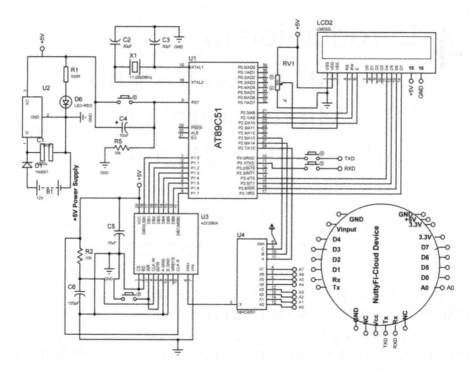

FIGURE 4.1   Schematics of customized with 8051 and NuttyFi board.

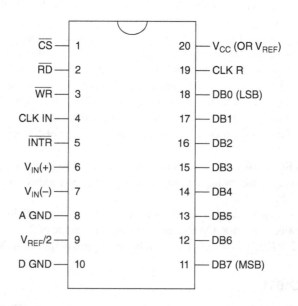

FIGURE 4.2   View of 0804.

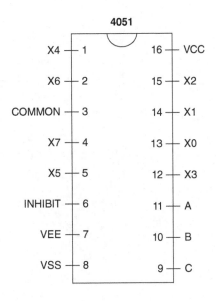

FIGURE 4.3   View of 4051.

inputs (A, B, and C), eight independent inputs/outputs (Xn), a common input/output (Z), and a digital enable input (E). When E is HIGH, the switches are turned off. Inputs include clamp diodes. This enables the use of current limiting resistors to interface inputs to voltages in excess of Vcc.

Pins 9, 10, and 11 are the three digital inputs that select which way the switch will be active. Depending on the value of these three bits, the common pin will be connected to any one of the various X pins. A is the least-significant bit, and C is the most-significant. So when A=B=C=0, common is connected to X0 and when A=B=1 and C=0, common is connected to X3.

The power supply for this chip is a little bit complicated, because it aims to support both negative and positive (analog) voltages through the switch. In "normal" use, you'd connect Vcc to +5V, VSS to -5V, and VEE to 0V, or the signal ground. What's important here is that the Vcc to VSS voltages span the input range, and VEE to Vcc is in the relevant range for our logic signals. Since we're passing single-sided (0V to 5V or 9V) square waves through the feedback path, and using the same voltage range for our logic signaling, we can connect both VSS and VEE to ground. Figure 4.3 shows the view of 4051.

4.1.3   NUTTYFI

NUTTYFI is an IoT hardware platform based on ESP8266 12e series that enables user to build IoT products, research analysis systems, automation, and projects. Using NUTTIFY, users can monitor, manage, control, and search devices from any part of the world. It can also interface with any IoT web servers, icloud, Local or IoT mobile platform, or application easily. Number of services open source web servers and mobile apps are available in the Internet that is freeware to use.

It has 8 digital pins, (as shown in Figure 4.4) 1 Analog pin-A0, V_{input}—From 5V to 21V, 3.3V output pin, UART Pins to flash program to NUTTYFI Cloud Device using FTDI UART Bridge.

4.1.4   RESET CIRCUIT

RESET is an active High input when RESET is set to High, 8051 goes back to the power on state. The 8051 is reset by holding the RST high for at least two machine cycles and then returning it low, as shown in Figure 4.5.

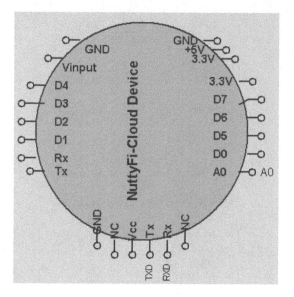

FIGURE 4.4   NuttyFi.

There are two methods of reset circuit:

1. Power On Reset
   Initially charging of capacitor makes RST High.
   When capacitor charges fully it blocks DC.
2. Manual Reset
   Closing the switch momentarily will make RST High.

4.1.5   CRYSTAL CIRCUIT

A crystal oscillator is an electronic oscillator circuit, which is used for the mechanical resonance of a vibrating crystal of piezoelectric material. It will create an electrical signal with a given frequency. This frequency is commonly used to keep track of time for example: wrist watches are used in digital integrated circuits to provide a stable clock signal and also used to stabilize frequencies for radio transmitters and receivers. Quartz crystal is mainly used in radio-frequency (RF) oscillators. Quartz crystal is the most common type of piezoelectric resonator; in oscillator circuits we

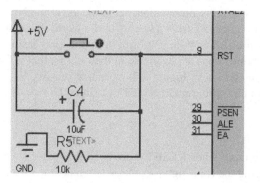

FIGURE 4.5   Reset circuit.

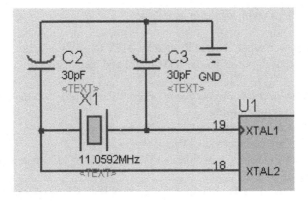

FIGURE 4.6 Crystal circuit.

are using them so it became known as crystal oscillators. Crystal oscillators must be designed to provide a load capacitance, as shown in Figure 4.6.

4.1.6 LCD CIRCUITRY

"Liquid Crystal Display" (LCD) could be a sort of show technology that produces use of liquid crystals that open or shut by an electrical current. These liquid crystals square measure the idea for alphanumeric display technology. The liquid crystals square measure manufactured from complicated molecules. Similar to water, they alter their state from solid to liquid, reckoning on the temperature to that they're exposed. Once in an exceedingly liquid state, the molecules move around however square measure possible to create a line in an exceedingly bound direction, permitting them to replicate light-weight. Crystals square measure organized in an exceedingly matrix with teams of three crystals of the colors red, inexperienced, and blue, forming a section referred to as a picture element. LCD has the distinct advantage of getting low-power consumption than the semiconductor diode, as shown in Figure 4.7.

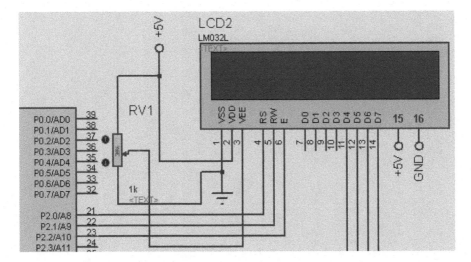

FIGURE 4.7 LCD interfacing with 8051.

The connections of LCD with 8051 are as follow:

1. Connect +12V power supply output to input of +12V to +5V convertor.
2. Connect +5V output to power up the 8051 board and also connect to other LCD to power up.
3. For programming, user can use hardware programmer to program the 8051 uC or one can also use in system programming.
4. Pins 1,16 of LCD are connected to GND of Power Supply.
5. Pins 2,15 of LCD are connected to +Vcc of Power Supply.
6. Two Fixed pins of POT are connected to +5V and GND of LCD and Variable lag of POT is connected to pin 3 of LCD.
7. RS, RW, and E pins of LCD are connected to pins P30, P31, and P32 of the 8051 board, respectively.
8. D4, D5, D6, and D7 pins of LCD are connected to pins P34, P35, P36, and P37 of the 8051 board, respectively.

Section B

Interfacing of 8051 Microcontroller and NuttyFi/ESP8266 with I/O devices

5 Interfacing of 8051 and NuttyFi/ESP8266 with LED

Light Emitting Diode (LED) is semiconductor source of illumination that emits light once current start flowing through it. Once an acceptable current is applied to the leads, negatrons area unit able to combine with electron holes and emits energy within the photons. The color of emitted energy is decided by the band gap of the semiconductor.

Most of the LEDs are made of a spread of inorganic materials of semiconductor. Table 5.1 shows the colors with different wavelength and material.

The V-I characteristic of a LED is analogous to a diode, which implies that a tiny low amendment in voltage will cause an outsized amendment in current. When applied voltage across the LED exceeds the forward fall by a low quantity, the present rating is also exceeded by an outsized quantity, which may damage the LED. The standard resolution is to use constant-current power to stay the LED below the most current rating. Figure 5.1 shows the LED circuit for limiting the current.

Figure 5.2 shows the block diagram of the system for LED interfacing. The system comprises of a +12V power supply, a +12V to +5V converter, a NuttyFi/8051 microcontroller, a resistor 330 Ohm, and the LEDs. The objective of the system is to blink the LEDs using ESP8266/NuttyFi/8051 microcontroller. Table 5.2 shows the list of components used to design the system.

5.1 CIRCUIT DIAGRAM

Description of interfacing of LED with ESP 8266

The interfacing of the devices is as follows:

1. Connect output of +12V power supply to input of +12V to +5V convertor.
2. Connect output of +5V power supply to +5V of the NuttyFi board.
3. Connect anode terminals of LEDs through 330 Ohm resistors to D1, D2 pins of the NuttyFi, respectively.
4. Cathode terminals of LEDs are connected to the ground.
5. Connect FTDI programmer to FTDI connector of the NuttyFi board.

Figure 5.3 shows the circuit diagram of the system.

Description of interfacing of 8051 with LED with 8051

The interfacing of the devices is as follows:

1. Connect output of +12V power supply to input of +12V to +5V convertor.
2. Connect output of +5V power supply to +5V of the 8051 board.
3. Connect anode terminals of LEDs through 330 Ohm resistors to P2.0, P2.1 pins of the 8051, respectively.
4. Cathode terminals of LEDs are connected to the ground.

Figure 5.4 shows the circuit diagram of the system.

TABLE 5.1
LED Color with Respect to Semiconductor Material

Color	Semiconductor Material	Wavelength [nm]	Voltage Drop [ΔV]
Infrared	Gallium arsenide (GaAs) Aluminum gallium arsenide (AlGaAs)	$\lambda > 760$	$\Delta V < 1.63$
Red	Aluminum gallium arsenide (AlGaAs) Gallium arsenide phosphide (GaAsP) phosphide (AlGaInP)	$610 < \lambda < 760$	$1.63 < \Delta V < 2.03$
Orange	Gallium arsenide phosphide (GaAsP) Aluminum gallium indium phosphide (AlGaInP)	$590 < \lambda < 610$	$2.03 < \Delta V < 2.10$
Yellow	Gallium arsenide phosphide (GaAsP) Aluminum gallium indium phosphide (AlGaInP)	$570 < \lambda < 590$	$2.10 < \Delta V < 2.18$
Green	Aluminum gallium indium phosphide (AlGaInP) Aluminum gallium phosphide (AlGaP) Indium gallium nitride (InGaN)	$500 < \lambda < 570$	$1.9^{[93]} < \Delta V < 4.0$
Blue	Zinc selenide (ZnSe) Indium gallium nitride (InGaN) Synthetic sapphire, Silicon carbide (SiC)	$450 < \lambda < 500$	$2.48 < \Delta V < 3.7$
Violet	Indium gallium nitride (InGaN)	$400 < \lambda < 450$	$2.76 < \Delta V < 4.0$
Ultraviolet	Indium gallium nitride (InGaN) (385–400 nm) Diamond (235 nm) Boron nitride (215 nm) Aluminum gallium nitride (AlGaN)	$\Lambda < 400$	$3 < \Delta V < 4.1$
Purple	Dual blue/red LEDs, blue with red phosphor, or white with purple plastic	Multiple types	$2.48 < \Delta V < 3.7$
White	**Cool/Pure White:** Blue/UV diode with yellow phosphor **Warm White:** Blue diode with orange phosphor	Broad spectrum	$2.8 < \Delta V < 4.2$

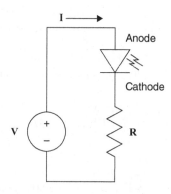

FIGURE 5.1   LED circuit for current limiting.

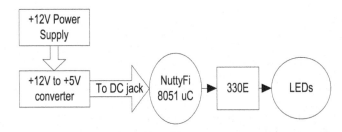

FIGURE 5.2 Block diagram of the interfacing of LED with ESP8266 and 8051.

TABLE 5.2
List of the Components Used to Design the System

Sr. No.	Name of the Components	Quantity	Specifications
	Components Used to Interface LED with ESP8266		
1	+12V power supply	01	Output- +12V/1A
2	+12V to 5V convertor	01	Output- +5V/1A
3	NuttyFi	01	Analog pin-1
			Digital pin-10
			FTDI connector for programmer
4	LED breakout board with 330 Ohm resistor	01	RED color LED as indicator
	Components Used to Interface LED with 8051		
1	+12V power supply	01	Output- +12V/1A
2	+12V to 5V convertor	01	To regulate the input voltage to +5V/1A
3	330 Ohm resistor	03	0.25 Watt
4	LED	03	RED
5	8051 development board	01	Atmel Make including crystal oscillator and reset circuit

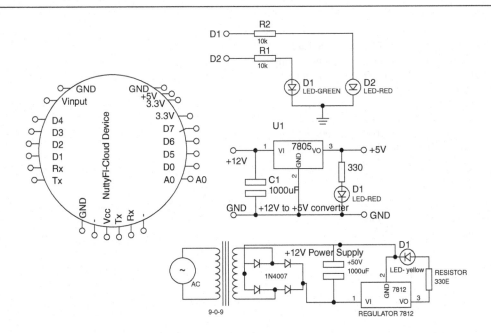

FIGURE 5.3 Circuit diagram for the interfacing of LED to NuttyFi.

5.2   PROGRAM

// Program to interface two LEDs with ESP8266/NuttyFi

```
int LED1_pin=D1; // assign variable to LED1
int LED2_pin=D2; // assign variable to LED1
void setup()
{
 pinMode(LED1_ pin, OUTPUT); // Assign pin1 to output pin
 pinMode(LED2_pin, OUTPUT); // Assign pin2 to output pin
}
void loop()
{
 digitalWrite(LED1_pin, HIGH); // make D1 pin HIGH
 digitalWrite(LED2_pin, HIGH); // make D2 pin HIGH
 delay(1000); // wait for 1 sec
 digitalWrite(LED1_pin, HIGH); // make D1 pin HIGH
 digitalWrite(LED2_pin, LOW); // make D2 pin LOW
 delay(1000); // wait for 1 sec
 digitalWrite(LED1_pin, LOW); // make D1 pin LOW
 digitalWrite(LED2_pin, HIGH); // make D2 pin HIGH
 delay(1000);  // wait for 1 sec
 digitalWrite(LED1_pin, LOW); // make D1 pin LOW
 digitalWrite(LED2_pin, LOW); // // make D2 pin LOW
 delay(1000); // wait for 1 sec
}
```

// Program to interface two LEDs with 8051

```
#include <REG52.h> // initializing of header file
void delay(); // declare of delay function

void main()//function: main         Objective: to glow LED
{
 while(1)
{
P20=1; //Switch OFF the LED connected at P20
delay(); //Random delay for some amount of time
P20=0; //Switch ON the LED connected at P20
delay();//Random delay for some amount of time

P21=1; //Switch OFF the LED connected at P21
delay();//Random delay for some amount of time
P21=0; //Switch ON the LED connected at P21
delay();//Random delay for some amount of time
}
}

//Function: delay          Objective: it provides delay to glow LED
void delay()
{
int i;
for(i=0;i<=100;i++);
 }
```

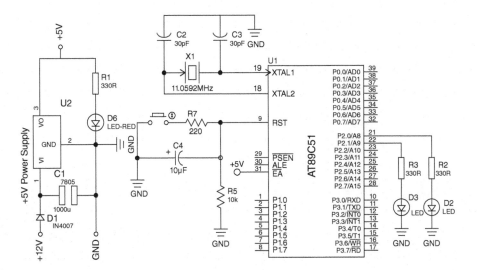

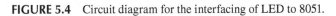

FIGURE 5.4   Circuit diagram for the interfacing of LED to 8051.

6 Interfacing of 8051 and NuttyFi/ESP8266 with Seven Segment Display

This chapter discusses the basic of seven segment display and its features. The interfacing of seven segment display with 8051 and NuttyFi is also discussed with the help of circuit diagram and programming.

6.1 SEVEN SEGMENT DISPLAY

A seven-segment display is an electronic device, which can be used to display decimal numeric values. Functional numerical displays (FND) can be employed in digital clocks, basic calculators, electronic meters, and different electronic devices that show numerical data.

Figure 6.1 shows the seven segment display. Every light-emitting diode has two connecting pins, one known as the "Anode" and other as the "Cathode". Seven segment displays can be categorized in two types—common cathode and common anode. Common anode has all the anode of the seven segments connected directly to power and the common cathode has all the cathode are connected to the ground.

1. Common cathode seven segment—In this type, cathode of the diodes are connected to logic "0" or ground. The individual segment measures a "HIGH" signal through a current limiting device to make it forward bias the individual anode terminal. Figure 6.2 shows the common cathode seven segment display.
2. Common anode seven segment—In this type, anode of the diodes are connected to logic "1". The individual segments measures a logic "LOW" signal through a current limiting device to make it forward bias the individual anode terminal. Figure 6.3 shows the common anode seven segment display.

Relying upon the digit to be displayed, the required set of LEDs is forward biased. As an example, to display the numerical digit zero, make LEDs placed at a, b, c, d, e, and f forward biased. The varied digits from zero through nine are showed employing a seven-segment display as shown, Figure 6.4.

To understand the working of seven segment display a system is designed. Figure 6.5 shows the block diagram of the system. The system comprises of a +12V power supply, a +12V to +5V converter, a NuttyFi/8051, a resistor 330 Ohm, and a seven-segment display. The objective of the project is to display the number on seven segment display using NuttyFi/8051.

Table 6.1 shows the list of the components used to design the system.

6.2 CIRCUIT DIAGRAM

Description of the interfacing of seven segment display with ESP8266/NuttyFi

The interfacing of the seven segment display is as follows:

1. Connect output of +12V power supply to input of +12V to +5V convertor.
2. Connect output of +5V power supply to +5V of the NuttyFi/ESP8266.
3. Connect input pins of seven segment (a–g and GND) to D1–D7 and GND pins of the NuttyFi respectively.
4. Connect FTDI programmer to FTDI connector of the NuttyFi board.

FIGURE 6.1   Seven segment display.

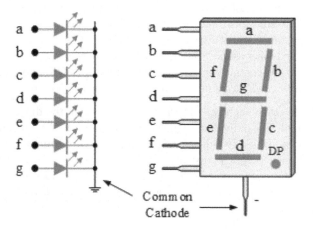

FIGURE 6.2   Common cathode seven segment display.

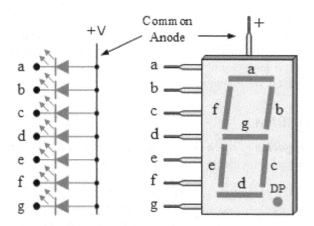

FIGURE 6.3   Common anode seven segment display.

FIGURE 6.4 Seven segment display for displaying numbers (0–9).

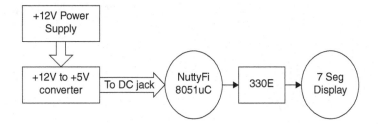

FIGURE 6.5 Block diagram of the system.

TABLE 6.1
Components List

Sr. No.	Name of the Components	Quantity	Specifications
	Components Used to Interface ESP8266 with LED		
1	+12V power supply	1	Output +12V/1A
2	+12V to +5V converter	1	Output +5V/1A
3	NuttyFi	1	Analog pin-1
			Digital pin-10
			FTDI connector for programmer
4	Seven segment breakout board with 330 Ohm resistor	1	Common anode RED color segments
	Components Used to Interface 8051 uC with LED		
1	+12V power supply	01	Output +12V/1A
2	+12V to 5V convert2r	01	To regulate the input voltage to +5V/1A
3	330 Ohm resistor	01	0.25 Watt
4	LED	03	RED
5	8051 development board	01	Atmel Make including crystal oscillator and reset circuit
6	Seven segment breakout with 330 Ohm resistor	1	Common anode RED color segments

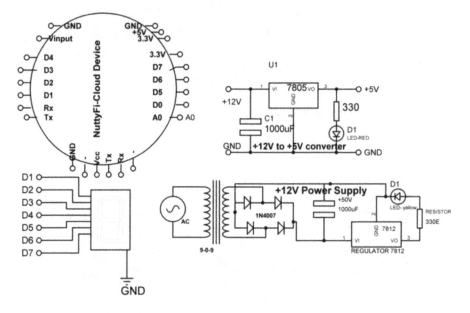

FIGURE 6.6   Circuit diagram for the interfacing of seven segment display with NuttyFi.

Figure 6.6 shows the circuit diagram of the system.

Description of interfacing of seven segment display with 8051

The interfacing of the seven segment display is as follows:

1. Connect output of +12V power supply to input of +12V to +5V converter.
2. Connect output of +5V power supply to +5V of 8051.
3. Connect input pins of seven segment (a–g and GND) to P3.0, P3.1, P3.2, P3.3, P3.4, P3.5, P3.6, P3.7, and GND pins of 8051 respectively.

Figure 6.7 shows the circuit diagram of the system.

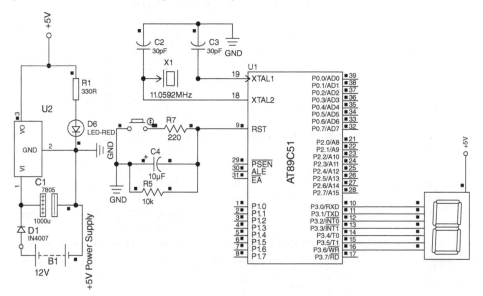

FIGURE 6.7   Circuit diagram for the interfacing of seven segment display with 8051.

6.3 PROGRAM

//Program to interface common anode seven segment with ESP8266

```
int A_pin=D1; // assign variable to D1 pin
int B_pin=D2; // assign variable to D2 pin
int C_pin=D3; // assign variable to D3 pin
int D_pin=D4; // assign variable to D4 pin
int E_pin=D5; // assign variable to D5 pin
int F_pin=D6; // assign variable to D6 pin
int G_pin=D7; // assign variable to D7 pin
void setup()
{
 pinMode(A_pin, OUTPUT);  // set D1 pin direction to output
 pinMode(B_pin, OUTPUT);  // set D2 pin direction to output
 pinMode(C_pin, OUTPUT);  // set D3 pin direction to output
 pinMode(D_pin, OUTPUT);  // set D4 pin direction to output
 pinMode(E_pin, OUTPUT);  // set D5 pin direction to output
 pinMode(F_pin, OUTPUT);  // set D6 pin direction to output
 pinMode(G_pin, OUTPUT);  // set D7 pin direction to output
}

void loop()
{
 ////// print 0
 digitalWrite(A_pin, HIGH);  // make D1 to HIGH
 digitalWrite(B_pin, HIGH);  // make D2 to HIGH
 digitalWrite(C_pin, HIGH);  // make D3 to HIGH
 digitalWrite(D_pin, HIGH);  // make D4 to HIGH
 digitalWrite(E_pin, HIGH);  // make D5 to HIGH
 digitalWrite(F_pin, HIGH);  // make D6 to HIGH
 digitalWrite(G_pin, LOW);  // make D7 to LOW
 delay(1000);            // wait for a second
 /// print 1
 digitalWrite(A_pin, LOW);  // make D1 to LOW
 digitalWrite(B_pin, HIGH);  // make D2 to HIGH
 digitalWrite(C_pin, HIGH);  // make D3 to HIGH
 digitalWrite(D_pin, LOW);  // make D4 to LOW
 digitalWrite(E_pin, LOW);  // make D5 to LOW
 digitalWrite(F_pin, LOW);  // make D6 to LOW
 digitalWrite(G_pin, LOW);  // make D7 to LOW
 delay(1000);            // wait for a second
 ///// print2
 digitalWrite(A_pin, HIGH);  // make D1 to HIGH
 digitalWrite(B_pin, HIGH);  // make D2 to HIGH
 digitalWrite(C_pin, LOW);  // make D3 to LOW
 digitalWrite(D_pin, HIGH);  // make D4 to HIGH
 digitalWrite(E_pin, HIGH);  // make D5 to HIGH
 digitalWrite(F_pin, LOW);  // make D6 to LOW
 digitalWrite(G_pin, HIGH);  // make D7 to HIGH
 delay(1000);            // wait for a second
}
```

//Program to interface common anode seven segment with 8051

```
#include<reg52.h>
void delay();
```

```
//function: main          Objective: to glow seven segment display
void main()
{
P3=0xff;
P2=0xff;
while(1)
{
P3=0xc0; // Command to display 0
delay(); // Some delay between the two digit display
P3=0xf9; // Command to display 1
delay(); // Some delay between the two digit display
P3=0xa4; // Command to display 2
delay(); // Some delay between the two digit display
P3=0xb0; // Command to display 3
delay(); // Some delay between the two digit display
P3=0x99; // Command to display 4
delay(); // Some delay between the two digit display
P3=0x92; // Command to display 5
delay(); // Some delay between the two digit display
P3=0x82; // Command to display 6
delay(); // Some delay between the two digit display
P3=0xf8; // Command to display 7
delay(); // Some delay between the two digit display
P3=0x80; // Command to display 8
delay(); // Some delay between the two digit display
P3=0x90; // Command to display 9
delay(); // Some delay between the two digit display
}
}
//Function: delay          Objective: it provides delay to glow seven segment display
void delay()
{
int i,j;
for(j=0;j<=100;j++)
{
for(i=0;i<=2000;i++)
{

}
}
}
```

7 Interfacing of 8051 and NuttyFi/ESP8266 with LCD

This chapter discusses the basic of liquid crystal display and its features. The interfacing of liquid crystal display with 8051 and NuttyFi is also discussed with the help of circuit diagram and programming.

7.1 INTRODUCTION OF LIQUID CRYSTAL DISPLAY

"Liquid Crystal Display" (LCD) is used to display alpha numeric characters. The liquid crystals matrixes are manufactured from complicated molecules. Similar to water, they alter their state from solid to liquid, reckoning on the temperature to that they're exposed. Once in an exceedingly liquid state, the molecules move around however square measure possible to create a line in an exceedingly bound direction, permitting them to replicate light-weight.

LCD has the distinct advantage of getting low power consumption than the semiconductor diode, as shown in Figure 7.1.

A liquid cell consists of a skinny layer (about 10 um) of a liquid sandwiched between two glass sheets with clear electrodes deposited on their within faces.

The Figure 7.2 shows the block diagram of the system. The system comprises of +12V power supply, +12V to +5V converter, NuttyFi, resistor 330 ohm, and LCD. The objective is to display the information on LCD using NuttyFi by writing the hex file in the flash memory. Table 7.1 shows the pin description of LCD.

Table 7.2 shows the list of the components used to design the system.

7.2 CIRCUIT DIAGRAM

Description of interfacing of ESP8266 with 16x2 LCD

Figure 7.3 shows the connection diagram of the system.

1. Connect +12V power supply output to input of +12V to +5V convertor.
2. Connect +5V output to power up the nuttyFi board and also connect to other peripheral to power up.
3. For programming uses—connect FTDI programmer to FTDI connector of the NuttyFi board.
4. Pins 1 to 16 of LCD are connected to GND.
5. Pins 2 to 15 of LCD are connected to +Vcc to Power Supply
6. Two fixed pins of variable resistor are connected to +5V and GND of LCD and variable pin is connected to pin 3 of LCD.
7. E, RS, and RW pins of LCD are connected to pins D1, GND, and D2 of the NuttyFi.
8. D4 to D7 pins of LCD are connected to pins D3, D4, D5, and D6 of the NuttyFi.

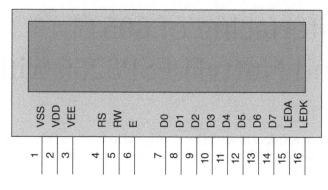

FIGURE 7.1   Pin out of LCD.

Description of interfacing of 8051 with 16x2 LCD

The interfacing of the 16x2 LCD with 8051 is as per the given guidelines. Figure 7.4 shows the connection diagram of the system.

1. Connect +12V power supply output to input of +12V to +5V convertor.
2. Connect +5V output to power up the 8051 board and also connect to other LCD to power up.
3. For programming user can use hardware programmer to program the 8051 uC or one can also use in system programming.
4. Pins 1, 16 of LCD are connected to GND of Power Supply.
5. Pins 2, 15 of LCD are connected to +Vcc of Power Supply.
6. Two fixed pins of POT are connected to +5V and GND of LCD and variable terminal of POT is connected to pin 3 of LCD.
7. RS, RW, and E pins of LCD are connected to pins P30, P31, and P32 of the 8051 board respectively.
8. D4, D5, D6, and D7 pins of LCD are connected to pins P34, P35, P36, and P37 of the 8051 board respectively.

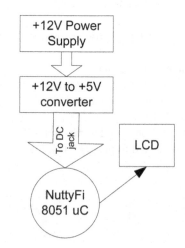

FIGURE 7.2   Block diagram of the system.

TABLE 7.1
Pin Description of LCD

Sr. No.	Pin No.	Pin Name	Pin Type	Pin Description	Pin Connection
1	1	Ground	Source Pin	This is a ground pin of LCD	Connected to the ground of the MCU/ Power source
2	2	Vcc	Source Pin	This is the supply voltage pin of LCD	Connected to the supply pin of Power source
3	3	V0/VEE	Control Pin	Adjusts the contrast of the LCD.	Connected to a variable POT that can source 0–5V
4	4	Register Select	Control Pin	Toggles between Command/ Data Register	Connected to a MCU pin and gets either 0 or 1 0 -> Command Mode 1-> Data Mode
5	5	Read/ Write	Control Pin	Toggles the LCD between Read/Write Operation	Connected to a MCU pin and gets either 0 or 1 0 -> Write Operation 1-> Read Operation
6	6	Enable	Control Pin	Must be held high to perform Read/Write Operation	Connected to MCU and always held high
7	7–14	Data Bits (0–7)	Data/Command Pin	Pins used to send Command or data to the LCD	<u>In 4-Wire Mode</u> Only 4 pins (0–3) is connected to MCU <u>In 8-Wire Mode</u> All 8 pins(0–7) are connected to MCU
8	15	LED Positive	LED Pin	Normal LED like operation to illuminate the LCD	Connected to +5V
9	16	LED Negative	LED Pin	Normal LED like operation to illuminate the LCD connected with GND	Connected to ground

TABLE 7.2
Components List

Sr. No.	Name of the Components	Quantity	Specifications
	Components Used to Interface ESP8266 with LCD		
1	+12V power supply	1	Output- +12V/1A
2	+12V to 5V convertor	1	Output- +5V/1A
3	NuttyFi	1	Analog pin-1 Digital pin-10 FTDI connector for programmer
4	LCD breakout board	1	4 Pins for +5V and 4 Pins for GND
	LCD 20*4	1	16 pins LCD with backlight pins
	Components Used to Interface 8051 uC with LCD		
1	+12V power supply	01	Output- +12V/1A
2	+12V to 5V convertor	01	To regulate the input voltage to +5V/1A
3	330 Ohm resistor	01	0.25 Watt
4	LED	02	RED
5	8051 development board	01	Atmel Make including crystal oscillator and reset circuit
6	LCD breakout board	1	4 Pins for +5V and 4 Pins for GND
7	LCD 20*4	1	16 pins LCD with backlight pins

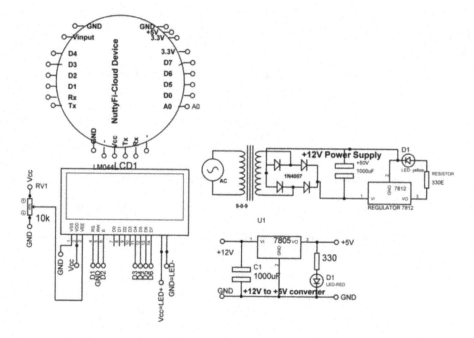

FIGURE 7.3   Circuit diagram of the system with NuttyFi.

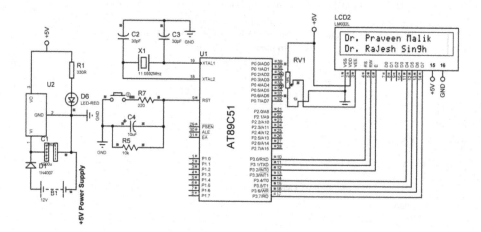

FIGURE 7.4   Circuit diagram of the system with 8051.

7.3   PROGRAM

Program to interface liquid crystal display (LCD) with ESP8266

```
#include <LiquidCrystal.h>
const int RS = D1, E = D2, D4 = D3, D5 = D4, D6 = D4, D7 = D5;// assign variable to LCD pins
LiquidCrystal lcd(RS, E, D4, D5, D6, D7);
void setup()
{
  lcd.begin(20, 4); // initialize LCD
lcd.setCursor(0, 0); // set cursor of the LCD
  lcd.print("DISPLAY SYSTEM"); // print string on LCD
```

```
  lcd.setCursor(0, 1); // set cursor of the LCD
  lcd.print("UsingLCD + NodeMCU"); // print string on LCD
}

void loop()
{
  lcd.clear(); // clear the previous contents of LCD
  lcd.setCursor(0, 2); // set cursor of the LCD
  lcd.print("Hi Gentleman"); // print string on LCD
  delay(2000); // wait for 2000 ms
  lcd.clear();// clear the previous contents of LCD
  lcd.setCursor(0, 2); // set cursor of the LCD
  lcd.print("what's UP"); // print string on LCD
  delay(2000); // wait for 2000 ms
}
```

Program to interface liquid crystal display (LCD) with 8051

```
#include<reg51.h>
sbit lcd_E  = P3^2; //Set P32 as enable pin
sbit lcd_RW = P3^1; //Set P31 as read write pin
sbit lcd_RS = P3^0; //Set P30 as resistor select pin

void lcd_init(); //LCD initialize command
void send_command(char command); // Command function
void send_data(char databyte); //Data function
void delay_lcd(); // LCD delay function
void display(char, char[]); //Display function
void delay(int);
void main() //Main function to display the contents
{
lcd_init(); //Initialize the LCD
delay(1); // Some delay for LCD
while(1) //Endless loop
{
display(0x80, "Praveen Malik"); //Display at first row of LCD
display(0xC0, "Rajesh Singh"); //Display at second row of LCD
}
}

void display(char x, char databyte[]) //Function to display at different row of LCD
{
char y=0;
send_command(x);
while(databyte[y]!='\0')
{
send_data(databyte[y]);
y++;
}
}

void lcd_init() //LCD initialize function
{
send_command(0x03);
delay_lcd();
send_command(0x28);
```

```
delay_lcd();
send_command(0x0E);
delay_lcd();
send_command(0x01);
delay_lcd();
send_command(0x0c);
delay_lcd();
send_command(0x06);
delay_lcd();
}

void send_command(char command)
{
char x,y;
lcd_RS = 0; /* (lcd_RS) P2.3---->RS , RS=0 for Instruction Write */
lcd_RW = 0; /* (lcd_RW) P2.2---->R/W, W=0 for write operation */
lcd_E = 1; /* (lcd_E)  P2.5---->E ,E=1 for Instruction execution */
x=command;
x=x >> 4;
P3=P3 & 0xf0;
P3=P3 | x;

lcd_E = 0;/* E=0; Disable lcd */
lcd_RW = 1; /* W=1;  Disable write signal */
lcd_RS = 1; /* RS=1 */

lcd_RS = 0;  /* (lcd_RS) P2.3---->RS , RS=0 for Instruction Write */
lcd_RW = 0; /* (lcd_RW) P2.2---->R/W, W=0 for write operation */
lcd_E = 1; /* (lcd_E)  P2.5---->E ,E=1 for Instruction execution */

y=command;
y=y & 0x0f;
P3=P3 & 0xf0;
P3=P3 | y;

lcd_E = 0; /* E=0; Disable lcd */
lcd_RW = 1; /* W=1;  Disable write signal */
lcd_RS = 1;  /* RS=1 */
}

void send_data(char databyte)
{
char x,y;
lcd_RS = 1; /* (lcd_RS) P2.3---->RS , RS=1 for Data Write */
lcd_RW = 0; /* (lcd_RW) P2.2---->R/W, W=0 for write operation */
lcd_E = 1; /* (lcd_E) P2.5---->E ,E=1 for Instruction execution */

x=databyte;
x=x >> 4;
P3=P3 & 0xf0;
P3=P3 | x;

lcd_E = 0; /* E=0; Disable lcd */
lcd_RW = 1; /* W=1; Disable write signal */
lcd_RS  = 0; /* RS=0 */
```

```
lcd_RS = 1;/* (lcd_RS) P2.3---->RS , RS=1 for Data Write */
lcd_RW = 0;/* (lcd_RW) P2.2---->R/W, W=0 for write operation */
lcd_E  = 1;/* (lcd_E) P2.5---->E ,E=1 for Instruction execution */

y=databyte;
y=y & 0x0f;
P3=P3 & 0xf0;
P3=P3 | y;

lcd_E = 0;/* E=0; Disable lcd */
lcd_RW = 1;/* W=1; Disable write signal */
lcd_RS  = 0;/* RS=0 */
}

void delay_lcd() //Some random delay for LCD
{
int j;
for(j=0;j<100;j++);
}

void delay(int x) //Some random delay
{
int i,j;
for(i=0;i<=x;i++)
{
for(j=0;j<=32000;j++);
}
```

8 Interfacing of 8051 and NuttyFi/ESP8266 with Analog Sensor

This chapter discusses the basic of analog sensors and its features. The interfacing of analog sensors with 8051 and NuttyFi is also discussed with the help of circuit diagram and programming.

Analog sensor converts the environmental parameters into electrical signal like voltage and current. The output voltage may be in the range of 0 to 5V.

There are different types of sensors, which produce continuous analog output signal. The signal produced by the analog sensor is proportional to the change in the environmental parameter.

8.1 POTENTIOMETER

A **potentiometer** (POT) is a three-terminal resistor with a sliding or rotating contact. It forms an adjustable voltage divider. The rotating terminal is a wiper and it helps POT to act as a *variable resistor* or *rheostat*. POTs can be used to control electrical devices such as volume controls on audio equipment. Figure 8.1 shows the POT.

Figure 8.2 shows the block diagram to interface POT with NuttyFi/8051. The system comprises +12V power supply, +12V to +5V converter, NuttyFi, LED with resistor 330 Ohm, POT, and LCD. The objective of the system is to display the levels generated by ADC on liquid crystal display (LCD) using NuttyFi. If the levels increased by certain level indicator will be "ON". Table 8.1 shows the list of the components to design the system.

8.1.1 CIRCUIT DIAGRAM

Description of interfacing of POT with ESP8266

1. Connect +12V power supply output to input of +12V to +5V convertor.
2. Connect +5V output to power up the NuttyFi board and also connect to other peripherals to power up.
3. For programming connect FTDI programmer to FTDI connector of the NuttyFi board.
4. Pins 1, 16 of LCD are connected to GND of power supply.
5. Pins 2, 15 of LCD are connected to +Vcc of power supply.
6. Two fixed terminals of POT are connected to +5V and GND of LCD and variable terminal of POT is connected to pin 3 of LCD.
7. RS, RW, and E pins of LCD are connected to pins D1, GND, and D2 of the NuttyFi respectively.
8. D4, D5, D6, and D7 pins of LCD are connected to pins D3, D4, D5, and D6 of the NuttyFi respectively.
9. Connect +Vcc, GND and OUT pin of the **POT** to +5V, GND and A0 pin of the NuttyFi.
10. Connect input and GND pins of the LED breakout board to D7 and GND pins of NuttyFi respectively.

FIGURE 8.1   Potentiometer.

Figure 8.3 shows the circuit diagram for interfacing of POT with NuttyFi/ESP8266.
Description of interfacing of POT with 8051

1. Connect +12V power supply output to input of +12V to +5V convertor.
2. Connect +5V output to power up the ADC 0804 board and also connect to LCD to power up.
3. LCD data lines are connected to the port of P3 of 8051 uC.
4. Control lines of LCD RS, RW, and E are connected to P2.0, P2.1, and P2.2 of 8051 respectively.
5. Connect +Vcc, GND and OUT pin of the **POT** to +5V, GND and input pins of the ADC 0804 respectively.
6. Connect 8 output pin of ADC 0804 to the P1 port of the 8051 uC.
7. For programming user can use hardware programmer to program the 8051 uC or one can also use in system programming.

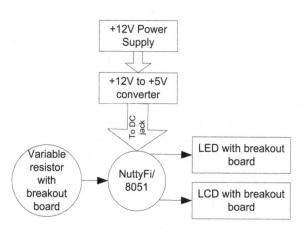

FIGURE 8.2   Block diagram to interface POT with NuttyFi.

TABLE 8.1
Components to Design the System

Sr. No.	Name of the Components	Quantity	Specifications
	Components Used to Interface ESP8266 with Variable Resistor		
1	+12V power supply	1	Output- +12V/1A
2	+12V to 5V convertor	1	Output- +5V/1A
3	NuttyFi	1	Analog pin-1
			Digital pin-10
			FTDI connector for programmer
4	LCD 20*4	1	16 pins LCD with backlight pins
5	POT with breakout board	1	Three pins
6	LED with breakout board	1	One RED color LED
	Components Used to Interface 8051 uC with Variable Resistor		
1	+12V power supply	01	Output- +12V/1A
2	+12V to 5V convertor	01	To regulate the input voltage to +5V/1A
3	330 Ohm resistor	01	0.25 Watt
4	LED	01	RED
5	8051 development board	01	Atmel Make including crystal oscillator and reset circuit
6	LCD 16x2	1	16 pins LCD with backlight pins
7	Variable resistor with breakout board	1	5K resistor
8	ADC 0804 breakout board	01	Single channel 8 bit

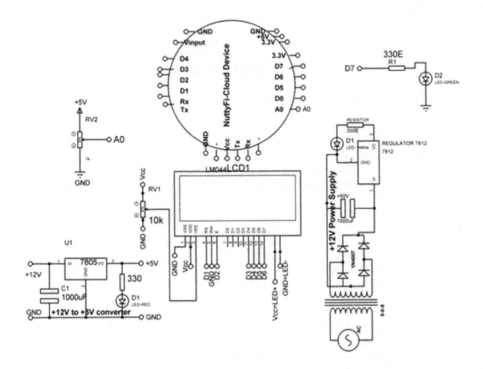

FIGURE 8.3 Circuit diagram for interfacing of POT with NuttyFi/ESP8266.

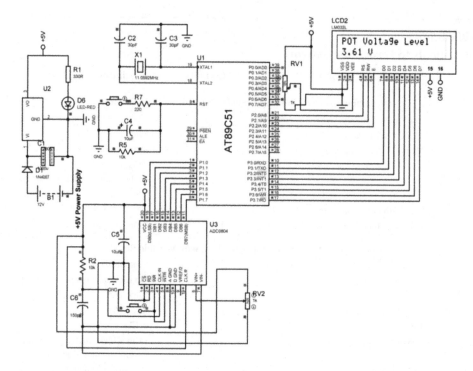

FIGURE 8.4 Circuit diagram for interfacing of POT with 8051.

Figure 8.4 shows the circuit diagram for interfacing of POT with 8051.

8.1.2 PROGRAM

Program to interface POT with ESP8266

```
#include <LiquidCrystal.h>
LiquidCrystal lcd(D1, D2, D3, D4, D5, D6); // add library of LCD

const int POT_Pin=A0;    // assign A0 pin to POT
const int INDICATOR_PIN=D7; // assign D7 as indicator pin
void setup()
{
 pinMode(INDICATOR_PIN, OUTPUT);     // set D7 pin as an output
 lcd.begin(20, 4); // initialize LCD
 lcd.print("POT Control..."); // print string on LCD
}

void loop()
{
  int POT_Pin_LEVEL = digitalRead(POT_Pin);// Read POT pin
  lcd.setCursor(0, 1); // set cursor of LCD
  lcd.print("ACTUAL_LEVEL:"); //print string on LCD
  if (POT_Pin_LEVEL >= 240) // compare level
  {
  lcd.setCursor(0, 3); // set cursor of LCD
  lcd.print("LEVEL_EXCEED    ");  // print string on LCD
  digitalWrite(INDICATOR_PIN, HIGH); // set D7 to HIGH
  delay(20); // wait for 20mSec
  }
```

```
else
 {
  lcd.setCursor(0, 3); // set cursor of LCD
  lcd.print("LEVEL NORMAL    "); // print string on LCD
  digitalWrite(INDICATOR_PIN, LOW); // set D7 to HIGH
  delay(20); // wait for 20mSec
 }
}
```

Program to interface POT with ADC0804 and 8051

```
#include<reg52.h>        //initializing of reg52 header file
#include<lcd.h>          //initializing of LCD header file
void disp_name();        //declaration of disp_name function
void delay();            //declaration of delay function
void disp_name1();       //declaration of disp_name1 function
char databyte[]={"0123456789"};
char a,b,c;

void main()     //definition of main function
{
P1=0xff;        //initialize the port 1 as 11111111
lcd_init();     //call the LCD initialize function
disp_name();    //call the disp_name function
lcd_init1();    //call the LCD1 initialize function
disp_name1();   //call the disp_name1 function

while(1)        //endless loop
{
i=P1; // read P1 port for getting analog sensor data
a=i%10; // add scaling factor
b=i/10; // add scaling factor
c=i/100; // add scaling factor
send_command(0xc0);         // send command to first location of LCD
send_data(databyte[c]);     //send first data
send_command(0xc1);         // send command to second location of LCD
send_data(databyte[b]);     //send second data
send_command(0xc2);         // send command to third location of LCD
send_data(databyte[a]);     //send third data
delay();                    //delay for some moment
}
}

void disp_name()            //function to display data on first line of LCD
{
char i=0;
char databyte[]={"POT Voltage Level "};      //Contents to be displayed
send_command(0x80);                          //Command of first line
while(databyte[i]!='\0')
{
send_data(databyte[i]);                      //send the data
i++;
}
}

void disp_name1()           //function to display data on second line of LCD
{
```

```
char i=0;
char databyte[]={"              "};        //contents to be displayed
send_command(0xC0);                        //command of second line
while(databyte[i]!='\0')
{
send_data(databyte[i]);                    //send the data
i++;
}
}
```

8.2 TEMPERATURE SENSOR

Temperature sensors are available in both forms—digital and analog. Commonly used analog temperature sensor is thermistor. For different applications, different types of thermistors are available. Thermistor detects change in temperature as a thermally sensitive resistor. If the temperature increases, then the electrical resistance of thermistor increases and resistance decreases as temperature decreases.

LM35 is an integrated analog temperature sensor whose electrical output is proportional to degree Centigrade. LM35 sensor does not require any external calibration. Figure 8.5 shows LM35.

Figure 8.6 shows the block diagram for interfacing of LM35 with NuttyFi/8051. The system comprises of +12V power supply, +12V to +5V converter, NuttyFi, 8051 board, LED with resistor 330 Ohm, LM35 with breakout board, and LCD. The objective of the system is to display the environmental temperature on LCD using NuttyFi. If the temperature exceed by certain level, indicator will be "ON". Table 8.2 shows the list of the components to design the system.

8.2.1 CIRCUIT DIAGRAM

Description of interfacing of LM35 with ESP8266

1. Connect +12V power supply output to input of +12V to +5V convertor.
2. Connect +5V output to power up the NuttyFi board and also connect to other peripherals to power up.
3. For programming, connect FTDI programmer to FTDI connector of the NuttyFi board.
4. Pins 1, 16 of LCD are connected to GND of power supply respectively.

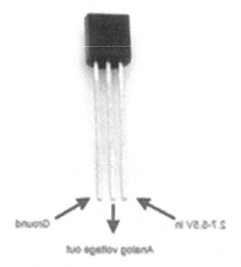

FIGURE 8.5 LM35 sensor.

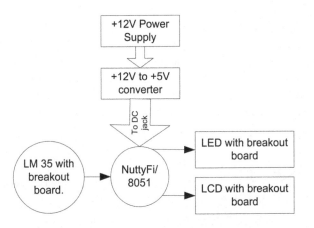

FIGURE 8.6   Block diagram for interfacing of LM35 with NuttyFi/8051.

5. Pins 2, 15 of LCD are connected to +Vcc of power supply respectively.
6. Two fixed terminals of POT are connected to +5V and GND of LCD and variable terminal of POT is connected to pin 3 of LCD.
7. RS, RW, and E pins of LCD are connected to pins D1, GND, and D2 of the NuttyFi respectively.
8. D4, D5, D6, and D7 pins of LCD are connected to pins D3, D4, D5, and D6 of the NuttyFi.
9. Connect +Vcc, GND and OUT pin of the **LM35** to +5V, GND and A0 pins of the NuttyFi respectively.
10. Connect input and GND pins of the LED breakout board to D7 and GND pins of NuttyFi respectively.

TABLE 8.2
Components Used to Design System

Sr. No.	Name of the Components	Quantity	Specifications
	Components Used to Interface ESP8266 with Variable Resistor		
1	+12V power supply	01	Output- +12V/1A
2	+12V to 5V convertor	01	Output- +5V/1A
3	NuttyFi	01	Analog pin-1
			Digital pin-10
			FTDI connector for programmer
4	LCD 20*4	01	16 pins LCD with backlight pins
5	POT with breakout board	01	Three pins
6	LED with breakout board	01	One RED color LED
7	LM35 with breakout board	01	Three pins
	Components Used to Interface 8051 uC with Variable Resistor		
1	+12V power supply	01	Output- +12V/1A
2	+12V to 5V convertor	01	To regulate the input voltage to +5V/1A
3	330 Ohm resistor	01	0.25 Watt
4	LED	01	RED
5	8051 development board	01	Atmel Make including crystal oscillator and reset circuit
6	LCD 16x2	01	16 pins LCD with backlight pins
7	LM35 with breakout board	01	Three pins
8	ADC 0804 breakout board	01	Single channel 8 bit

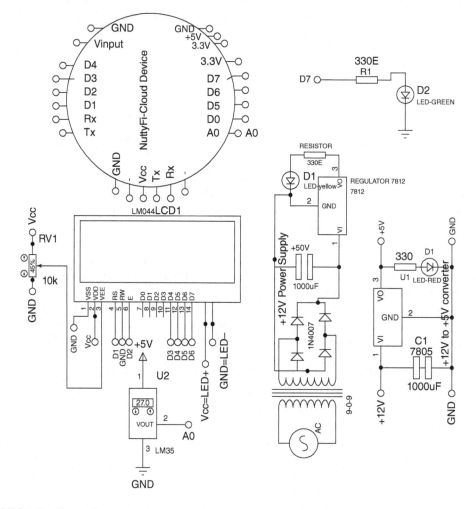

FIGURE 8.7   Circuit diagram for interfacing of LM35 with NuttyFi/ESP8266.

Figure 8.7 shows the circuit diagram for interfacing of LM35 with NuttyFi/ESP8266.
Description of interfacing of LM35 temperature sensor with 8051

1. Connect +12V power supply output to input of +12V to +5V convertor.
2. Connect +5V output to power up the ADC 0804 board and also connect to LCD to power up.
3. LCD data lines are connected to the port of P3 of 8051 uC.
4. Control lines of LCD RS, RW, and E are connected to P2.0, P2.1, and P2.2 respectively.
5. Connect +Vcc, GND and OUT pin of the **LM35** to +5V, GND and input pins of the ADC 0804 respectively.
6. Connect 8 output pin of ADC 0804 to the P1 port of the 8051 uC.
7. For programming, user can use hardware programmer to program the 8051 uC or one can also use in system programming.

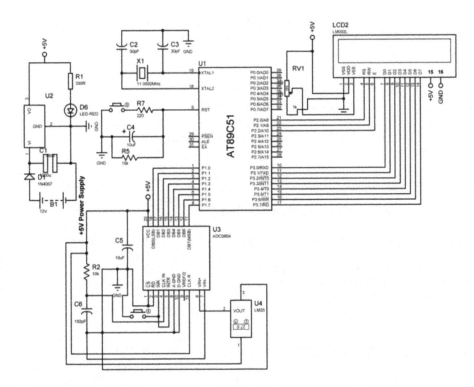

FIGURE 8.8 Circuit diagram for interfacing of LM35 with 8051.

Figure 8.8 shows the circuit diagram for interfacing of LM35 with 8051.

8.2.2 PROGRAM

Program to interface POT with ESP8266

```
#include <LiquidCrystal.h>
LiquidCrystal lcd(D1, D2, D3, D4, D5, D6); // add library of LCD
const int LM35Sensor_Pin=A0;    // assign A0 to LM35 data pin
const int INDICATOR_PIN=D7; // assign D7 to indicator pin
void setup()
{
 pinMode(INDICATOR_PIN, OUTPUT);    // set D7 as an output pin
 lcd.begin(20, 4); // initialize LCD
 lcd.print("TEMP Monitoring..."); // print string on LCD
}

void loop()
{
  int LM35SENSOR_Pin_LEVEL = digitalRead(LM35Sensor_Pin);// Read LM35 Sensor pin
  int TEMP_ACTUAL=LM35SENSOR_Pin_LEVEL/2; // scale level
  lcd.setCursor(0, 1); // set cursor on LCD
  lcd.print("ACTUAL_LEVEL:"); // print string on LCD
  lcd.setCursor(0, 2); // set cursor on LCD
  lcd.print(TEMP_ACTUAL); // print integer on LCD
 if (LDRSENSOR_Pin_LEVEL >= 40) // compare level
 {
  lcd.setCursor(0, 3); // set cursor on LCD
```

```
  lcd.print("TEMP_EXCEED    "); // print string on LCD
  digitalWrite(INDICATOR_PIN, HIGH); // make indicator pin HIGH
  delay(20); // wait for 20mSec
 }
else
 {
  lcd.setCursor(0, 3); // set cursor on LCD
  lcd.print("TEMP NORMAL    "); // print string on LCD
  digitalWrite(INDICATOR_PIN, LOW); // make indicator pin HIGH
  delay(20); // wait for 20mSec
 }
}
```

Program to interface POT with ADC0804 and 8051

```
#include<reg52.h>    //initializing of reg52 header file
#include<lcd.h>      //initializing of LCD header file
void disp_name();    //declaration of disp_name function
void delay();        //declaration of delay function
void disp_name1();   //declaration of disp_name1 function
char databyte[]={"0123456789"};
char a,b,c;

void main()      //definition of main function
{
P1=0xff;        //initialize the port 1 as 11111111
lcd_init();     //call the LCD initialize function
disp_name();    //call the disp_name function
lcd_init1();    //call the LCD1 initialize function
disp_name1();   //call the disp_name1 function

while(1)        //endless loop
{
i=P1; // read P1 port for getting analog sensor data
a=i%10; // add scaling factor
b=i/10; // add scaling factor
c=i/100; // add scaling factor
send_command(0xc0);          // send command to first location of LCD
send_data(databyte[c]);      //send first data
send_command(0xc1);          // send command to second location of LCD
send_data(databyte[b]);      //send second data
send_command(0xc2);          // send command to third location of LCD
send_data(databyte[a]);      //send third data
delay();                     //delay for some moment
}
}

void disp_name()              //function to display data on first line of LCD
{
char i=0;
char databyte[]={"Temperature reading is "};        //Contents to be displayed
send_command(0x80);                                 //Command of first line
while(databyte[i]!='\0')
{
send_data(databyte[i]);                        //send the data
i++;
```

```
}
}

void disp_name1()              //function to display data on second line of LCD
{
char i=0;
char databyte[]={"    Degree cent"};       //contents to be displayed
send_command(0xC0);                         //command of second line
while(databyte[i]!='\0')
{
send_data(databyte[i]);              //send the data
i++;
}
}
```

8.3 LIGHT DEPENDENT RESISTOR

Light dependent resistor (LDR) is an analog sensor, which can detect light intensity of a light source, also known as light intensity sensor. The sensors can be classified into different types—photoresistor, Cadmium Sulfide (CdS), and photocell. LDR can be used to switch on and off loads automatically. The resistance of the LDR increases with decrease in light and decreases with increase in light. Figure 8.9 shows the LDR sensor.

Figure 8.10 shows the block diagram for interfacing of LDR with NuttyFi/8051. The system comprises of +12V power supply, +12V to +5V converter, NuttyFi, LED with resistor 330 Ohm, LDR

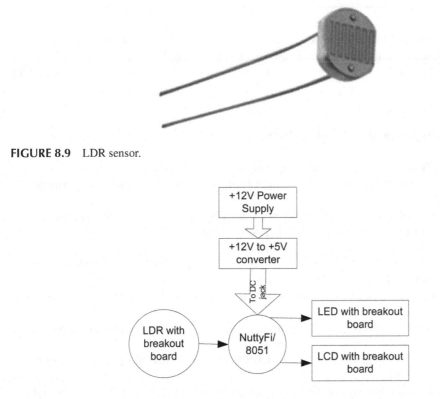

FIGURE 8.9 LDR sensor.

FIGURE 8.10 Block diagram for interfacing of LDR with NuttyFi/ESP8266.

TABLE 8.3

Components Used to Design the System

Sr. No.	Name of the Components	Quantity	Specifications
	Components Used to Interface ESP8266 with Variable Resistor		
1	+12V power supply	01	Output- +12V/1A
2	+12V to 5V convertor	01	Output- +5V/1A
3	NuttyFi	01	Analog pin-1
			Digital pin-10
			FTDI connector for programmer
4	LCD 20*4	01	16 pins LCD with backlight pins
5	LCD breakout board	1	4 Pins for +5V and 4 Pins for GND
6	LED with breakout board	01	One RED color LED
7	LDR with breakout board	1	Three pins
	Components Used to Interface 8051 uC with Variable Resistor		
1	+12V power supply	01	Output- +12V/1A
2	+12V to 5V convertor	01	To regulate the input voltage to +5V/1A
3	330 Ohm resistor	01	0.25 Watt
4	LED	01	RED
5	8051 development board	01	Atmel Make including crystal oscillator and reset circuit
6	LCD 16x2	01	16 pins LCD with backlight pins
7	LDR with breakout board	1	Three pins
8	ADC 0804 breakout board	01	Single channel 8 bit

with breakout board, and LCD. The objective of the system is to display the light intensity levels generate by LDR sensor using ADC on LCD using NuttyFi. If the light intensity exceeds by certain level, LED indicator will be "ON". Table 8.3 shows the list of the components to design the system.

8.3.1 Circuit Diagram

Description of interfacing of LDR with ESP8266

1. Connect +12V power supply output to input of +12V to +5V convertor.
2. Connect +5V output to power up the NuttyFi board and also connect to other peripheral to power up.
3. For programming, connect FTDI programmer to FTDI connector of the NuttyFi board.
4. Pins 1, 16 of LCD are connected to GND of power supply.
5. Pins 2, 15 of LCD are connected to +Vcc of power supply.
6. Two fixed terminals of POT are connected to +5V and GND of LCD and variable terminal of POT is connected to pin 3 of LCD.
7. RS, RW, and E pins of LCD are connected to pins D1, GND, and D2 of the NuttyFi respectively.
8. D4, D5, D6, and D7 pins of LCD are connected to pins D3, D4, D5, and D6 of the NuttyFi respectively.
9. Connect +Vcc, GND and OUT pin of the **LDR sensor** to +5V, GND and A0 pins of the NuttyFi respectively.
10. Connect input and GND pins of the LED breakout board to D7 and GND pins of NuttyFi respectively.

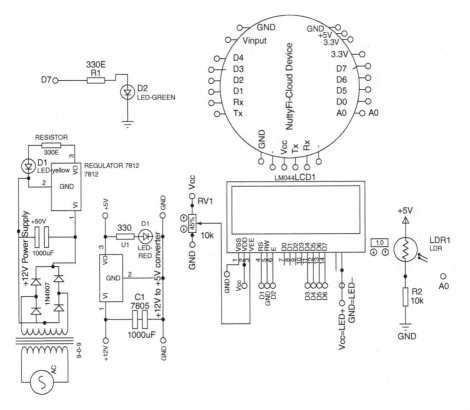

FIGURE 8.11 Circuit diagram for interfacing of LDR with ESP8266.

Figure 8.11 shows the circuit diagram for interfacing of LDR with ESP8266.
Description of interfacing of LDR with 8051

1. Connect +12V power supply output to input of +12V to +5V convertor.
2. Connect +5V output to power up the ADC 0804 board and also connect to LCD to power up.
3. LCD data lines are connected to the port of P3 of 8051 uC.
4. Control lines of LCD RS, RW, and E are connected to P2.0, P2.1, and P2.2 respectively.
5. Connect LDR as per the circuit diagram to the input pins of the ADC 0804.
6. Connect 8 output pin of ADC 0804 to the P1 port of the 8051 uC.
7. For programming user can use hardware programmer to program the 8051 uC or one can also use in system programming.

Figure 8.12 shows circuit diagram for interfacing of LDR with 8051.

8.3.2 Program

Program to interface LDR with ESP8266

```
#include <LiquidCrystal.h>
LiquidCrystal lcd(D1, D2, D3, D4, D5, D6); // add LCD library
const int LDRSensor_Pin=A0;   // assign A0 as LDR pin
const int LED_PIN=D7; // assign D7 as LED pin
```

```
void setup()
{
  pinMode(LED_PIN, OUTPUT);     // set D7 pin as an OUTPUT
  lcd.begin(20, 4); // initialize LCD
  lcd.print("LDR monitoring..."); // print string on LCD
}
void loop()
{
  int LDR_Pin_LEVEL = digitalRead(LDRSensor_Pin);// Read Fire Sensor pin
  lcd.setCursor(0, 1); // set cursor of LCD
  lcd.print("ACTUAL_LEVEL:"); // print string on LCD
  lcd.setCursor(0, 2);  // set cursor of LCD
  lcd.print(LDR_Pin_LEVEL);// print integer on LCD
  if (LDR_Pin_LEVEL >= 100) // compare the levels
  {
    lcd.setCursor(0, 3); // set cursor of LCD
    lcd.print("LEVEL_EXCEED   ");  // print string on LCD
    digitalWrite(LED_PIN, HIGH); // set D7 to HIGH
    delay(20); // wait for 20mSec
  }
  else
  {
    lcd.setCursor(0, 3); // set cursor of LCD
    lcd.print("LEVEL NORMAL   "); // print string on LCD
    digitalWrite(LED_PIN, LOW); // set D7 to HIGH
    delay(20); // wait for 20mSec
  }
}
```

Program to interface LDR with ADC0804 and 8051

```
#include<reg52.h>        //initializing of reg52 header file
#include<lcd.h>          //initializing of LCD header file
void disp_name();        //declaration of disp_name function
void delay();            //declaration of delay function
void disp_name1();       //declaration of disp_name1 function
char databyte[]={"0123456789"};
char a,b,c;

void main()       //definition of main function
{
P1=0xff;          //initialize the port 1 as 11111111
lcd_init();       //call the LCD initialize function
disp_name();      //call the disp_name function
lcd_init1();      //call the LCD1 initialize function
disp_name1();     //call the disp_name1 function
while(1)          //endless loop
{
i=P1; // read P1 port for getting analog sensor data
a=i%10; // add scaling factor
b=i/10; // add scaling factor
c=i/100; // add scaling factor
send_command(0xc0);        // send command to first location of LCD
send_data(databyte[c]);    //send first data
send_command(0xc1);        // send command to second location of LCD
send_data(databyte[b]);    //send second data
```

```
send_command(0xc2);        // send command to third location of LCD
send_data(databyte[a]);    //send third data
delay();                   //delay for some moment
}
}


void disp_name()           //function to display data on first line of LCD
{
char i=0;
char databyte[]={"Intensity Level      "};    //Contents to be displayed
send_command(0x80);                           //Command of first line
while(databyte[i]!='\0')
{
send_data(databyte[i]);                       //send the data
i++;
}
}


void disp_name1()          //function to display data on second line of LCD
{
char i=0;
char databyte[]={"   level (0-255)"};    //contents to be displayed
send_command(0xC0);                      //command of second line
while(databyte[i]!='\0')
{
send_data(databyte[i]);                  //send the data
i++;
}
}
```

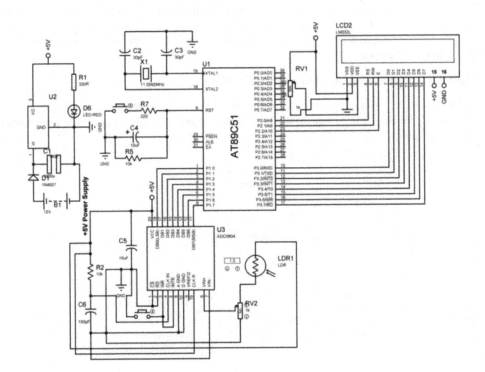

FIGURE 8.12 Circuit diagram for interfacing of LDR with 8051.

FIGURE 8.13 Flex sensor.

8.4 FLEX SENSOR

The flex sensor is a variable resistor where the resistance of the sensor changes with respect to the bend of the sensor. This sensor can be used as door sensor, robot whisker sensor, a sentient stuffed animal, etc. Figure 8.13 shows the flex sensor.

One side of the sensor is printed with a polymer ink with a conductive particles embedded in it. When the sensor is straight, the particles give the ink a resistance of about 30k Ohm. When the sensor is bent away from the ink, the conductive particles move further apart, increasing this resistance (to about 50k–70k Ohm when the sensor is bent to 90°). By measuring the resistance, the bent angle of sensor can be calculated.

Figure 8.14 shows the block diagram for interfacing of flex sensor with NuttyFi/8051. The system comprises of +12V power supply, +12V to +5V converter, NuttyFi, LED with resistor 330 Ohm, flex sensor with breakout board, and LCD. The objective of the system is to display the bent levels generated by flex sensor using ADC on LCD using NuttyFi. If the bent exceed to a certain level, buzzer indicator will be "ON". Table 8.4 shows the list of the components to design the system.

8.4.1 CIRCUIT DIAGRAM

Description of interfacing of flex sensor with ESP8266

1. Connect +12V power supply output to input of +12V to +5V convertor.
2. Connect +5V output to power up the NuttyFi board and also connect to other peripherals to power up.

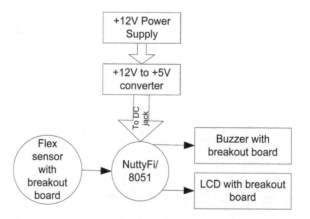

FIGURE 8.14 Block diagram for interfacing of flex sensor with NuttyFi/8051.

TABLE 8.4

Components Used to Design the System

Sr. No.	Name of the Components	Quantity	Specifications
	Components Used to Interface ESP8266 with Variable Resistor		
1	+12V power supply	01	Output- +12V/1A
2	+12V to 5V convertor	01	Output- +5V/1A
3	NuttyFi	01	Analog pin-1 Digital pin-10 FTDI connector for programmer
4	LCD 20*4	01	16 pins LCD with backlight pins
5	LED with breakout board	01	One RED color LED
6	Flex sensor with breakout board	1	Three pins
7	Buzzer with breakout board	1	+5V operated buzzer
	Components Used to Interface 8051 uC with Variable Resistor		
1	+12V power supply	01	Output- +12V/1A
2	+12V to 5V convertor	01	To regulate the input voltage to +5V/1A
3	330 Ohm resistor	01	0.25 Watt
4	LED	01	RED
5	8051 development board	01	Atmel Make including crystal oscillator and reset circuit
6	LCD 16x2	01	16 pins LCD with backlight pins
7	ADC 0804 breakout board	01	Single channel 8 bit
8	Flex sensor with breakout board	1	Three pins
9	Buzzer with breakout board	1	+5V operated buzzer

3. For programming, connect FTDI programmer to FTDI connector of the NuttyFi board.
4. Pins 1, 16 of LCD are connected to GND of power supply.
5. Pins 2, 15 of LCD are connected to +Vcc of power supply.
6. Two fixed terminals of POT are connected to +5V and GND of LCD and variable terminal of POT is connected to pin 3 of LCD.
7. RS, RW, and E pins of LCD are connected to pins D1, GND, and D2 of the NuttyFi respectively.
8. D4, D5, D6, and D7 pins of LCD are connected to pins D3, D4, D5, and D6 of the NuttyFi respectively.
9. Connect +Vcc, GND and OUT pin of the **Flex sensor** to +5V, GND and A0 pins of the NuttyFi respectively.
10. Connect input and GND pins of the buzzer breakout board to D7 and GND pins of NuttyFi.

Figure 8.15 shows the circuit diagram for interfacing of flex sensor with ESP8266.

Description of interfacing of flex sensor with 8051

1. Connect +12V power supply output to input of +12V to +5V convertor.
2. Connect +5V output to power up the ADC 0804 board and also connect to LCD to power up.
3. LCD data lines are connected to the port of P3 of 8051 uC.
4. Control lines of LCD RS, RW, and E are connected to P2.0, P2.1, and P2.2 respectively.
5. Connect +Vcc, GND and OUT pin of the flex sensor to +5V, GND and input pins of the ADC 0804, as shown in the circuit diagram.
6. Connect 8 output pin of ADC 0804 to the P1 port of the 8051 uC.

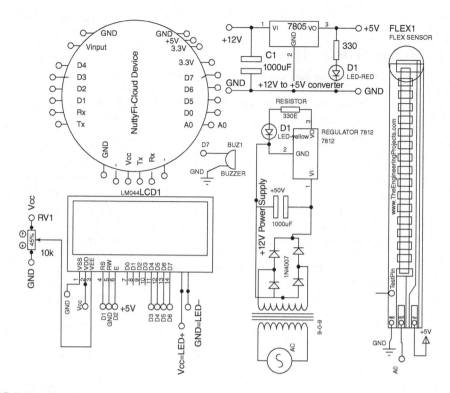

FIGURE 8.15 Circuit diagram for interfacing of flex sensor with ESP8266.

7. For programming user can use hardware programmer to program the 8051 uC or one can also use in system programming.
8. Connect the buzzer to P2.7 of 8051 uC.

Figure 8.16 shows the circuit diagram for interfacing of flex sensor with 8051.

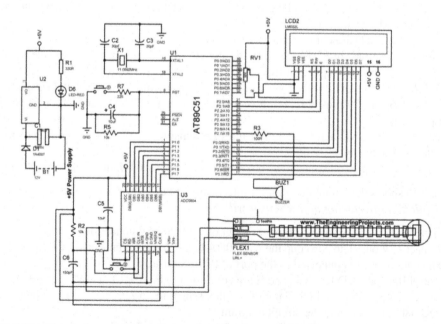

FIGURE 8.16 Circuit diagram for interfacing of flex sensor with 8051.

8.4.2 PROGRAM

Program to interface flex sensor with ESP8266

```
#include <LiquidCrystal.h>
LiquidCrystal lcd(D1, D2, D3, D4, D5, D6); // add LCD library
const int FLEXSensor_Pin=A0;    // assign A0 as flex sensor pin
const int BUZZER_PIN=D7; // assign D7 as buzzer pin
void setup()
{
  pinMode(LED_PIN, OUTPUT);   // set D7 as an output
  lcd.begin(20, 4); // initialize LCD
  lcd.print("Bent monitoring..."); // print string on LCD
}

void loop()
{
  int FLEX_Pin_LEVEL = digitalRead(FLEXSensor_Pin);// Read Fire Sensor pin
  int FLEX_SCALING = FLEX_Pin_LEVEL/4; // scale parameters
  lcd.setCursor(0, 1); // set cursor on LCD
  lcd.print("ACTUAL_LEVEL:"); // print string on LCD
  lcd.setCursor(0, 2); // set cursor on LCD
  lcd.print(FLEX_SCALING); // print integer on LCD
  if (FLEX_SCALING >= 50)  // compare scale factor
  {
   lcd.setCursor(0, 3); // set cursor on LCD
   lcd.print("bent exceed    "); // print string on LCD
   digitalWrite(BUZZER_PIN, HIGH); // make D7 pin to HIGH
   delay(20); // wait for 20mSec
  }
  else
  {
   lcd.setCursor(0, 3); // set cursor on LCD
   lcd.print("No bent   "); // print string on LCD
   digitalWrite(BUZZER_PIN, LOW); // make D7 pin to HIGH
   delay(20); // wait for 20mSec
  }
}
```

Program to interface flex sensor with ADC0804 and 8051

```
#include<reg52.h>        //initializing of reg52 header file
#include<lcd.h>          //initializing of LCD header file
void disp_name();        //declaration of disp_name function
void delay();            //declaration of delay function
void disp_name1();       //declaration of disp_name1 function
char databyte[]={"0123456789"};
char a,b,c;

void main()      //definition of main function
{
P1=0xff;         //initialize the port 1 as 11111111
lcd_init();      //call the LCD initialize function
disp_name();     //call the disp_name function
```

```
lcd_init1();      //call the LCD1 initialize function
disp_name1();     //call the disp_name1 function

while(1)          //endless loop
{
i=P1; // read P1 port for getting analog sensor data
a=i%10; // add scaling factor
b=i/10; // add scaling factor
c=i/100; // add scaling factor
send_command(0xc0);        // send command to first location of LCD
send_data(databyte[c]);    //send first data
send_command(0xc1);        // send command to second location of LCD
send_data(databyte[b]);    //send second data
send_command(0xc2);        // send command to third location of LCD
send_data(databyte[a]);    //send third data
delay();                   //delay for some moment
}
}

void disp_name()              //function to display data on first line of LCD
{
char i=0;
char databyte[]={"Actual Level      "};      //Contents to be displayed
send_command(0x80);                          //Command of first line
while(databyte[i]!='\0')
{
send_data(databyte[i]);                      //send the data
i++;
}
}

void disp_name1()             //function to display data on second line of LCD
{
char i=0;
char databyte[]={"    Flex scaling"};        //contents to be displayed
send_command(0xC0);                          //command of second line
while(databyte[i]!='\0')
{
send_data(databyte[i]);                      //send the data
i++;
}
}
```

8.5 GAS SENSOR

A gas sensor detects the presence of gases in an area for the safety. This type of device can be used to detect a gas leak or other emissions. It can be interfaced with a control system to control it automatically as per condition. Figure 8.17 shows gas sensor.

Figure 8.18 shows the block diagram for interfacing of gas sensor with NuttyFi/8051. The system comprises of +12V power supply, +12V to +5V converter, NuttyFi, LED with resistor 330 Ohm, gas sensor with breakout board, and LCD. The objective of the project is display the gas levels and PPM generated by gas sensor using ADC on LCD using NuttyFi by writing the hex file in the flash memory. If the gas exceed by certain level, hooter will "ON" to show the status. Table 8.5 shows the list of the components to design the system.

FIGURE 8.17   Gas sensor.

8.5.1   Circuit Diagram

Description of interfacing of ESP8266 with gas sensor

1. Connect +12V power supply output to input of +12V to +5V convertor.
2. Connect +5V output to power up the NuttyFi board and also connect to other peripheral to power up.
3. For programming, connect FTDI programmer to FTDI connector of the NuttyFi board.
4. Pins 1, 16 of LCD are connected to GND of power supply.
5. Pins 2, 15 of LCD are connected to +Vcc of power supply.
6. Two fixed terminals of POT are connected to +5V and GND of LCD and variable terminal of POT is connected to pin 3 of LCD.
7. RS, RW, and E pins of LCD are connected to pins D1, GND, and D2 of the NuttyFi respectively.
8. D4, D5, D6, and D7 pins of LCD are connected to pins D3, D4, D5, and D6 of the NuttyFi respectively.
9. Connect +Vcc, GND and OUT pin of the **gas sensor** to +5V, GND and A0 pins of the NuttyFi respectively.
10. Connect input, +12V and GND pins of the relay breakout board to D7, +12V and GND pins of NuttyFi.

Figure 8.19 shows the circuit diagram for interfacing of gas sensor with NuttyFi.

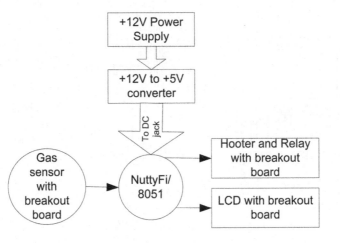

FIGURE 8.18   Block diagram for interfacing of gas sensor with NuttyFi.

TABLE 8.5
Components Used to Design the System

Sr. No.	Name of the Components	Quantity	Specifications
	Components Used to Interface ESP8266 with Gas Sensor		
1	+12V power supply	01	Output- +12V/1A
2	+12V to 5V convertor	01	Output- +5V/1A
3	NuttyFi	01	Analog pin-1 Digital pin-10 FTDI connector for programmer
4	LCD 20*4 break out board	01	16 pins LCD with backlight pins
5	LED with breakout board	01	One RED color LED
6	Gas sensor with breakout board	1	Three pins
7	Relay with breakout board	1	+5V operated buzzer
	Components Used to Interface 8051 uC with Gas Sensor		
1	+12V power supply	01	Output- +12V/1A
2	+12V to 5V convertor	01	To regulate the input voltage to +5V/1A
3	330 Ohm resistor	01	0.25 Watt
4	LED	01	RED
5	8051 development board	01	Atmel Make including crystal oscillator and reset circuit
6	LCD 16x2	01	16 pins LCD with backlight pins
7	ADC 0804 breakout board	01	Single channel 8 bit
8	Buzzer with breakout board	1	+5V operated buzzer
9	Gas sensor with breakout board	1	Three pins

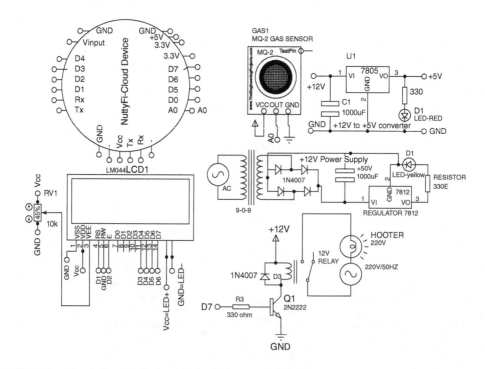

FIGURE 8.19 Circuit diagram for interfacing of gas sensor with NuttyFi.

Description of interfacing of gas sensor with 8051

1. Connect +12V power supply output to input of +12V to +5V convertor.
2. Connect +5V output to power up the ADC 0804 board and also connect to LCD to power up.
3. LCD data lines are connected to the port of P3 of 8051 uC.
4. Control lines of LCD RS, RW, and E are connected to P2.0, P2.1, and P2.2 respectively.
5. Connect +Vcc, GND and OUT pin of the gas sensor to +5V, GND and input pins of the ADC 0804.
6. Connect 8 output pin of ADC 0804 to the P1 port of the 8051 uC.
7. For programming user can use hardware programmer to program the 8051 uC or one can also use in system programming.

8.5.2 PROGRAM

Program to interface gas sensor with ESP8266

```
#include <LiquidCrystal.h>
LiquidCrystal lcd(D1, D2, D3, D4, D5, D6); // add library of LCD
const int GASSensor_Pin=A0;   // assign A0 pin to gas sensor
const int HOOTER_PIN=D7; // assign D7 to hooter
void setup()
{
 pinMode(LED_PIN, OUTPUT); // set D7 as an output
 lcd.begin(20, 4); // initialize LCD
 lcd.print("Gas Contents"); // print string on LCD
}

void loop()
{
  int GAS_Pin_LEVEL = digitalRead(GASSensor_Pin);// Read GAS Sensor pin
  int GAS_PPM = FLEX_Pin_LEVEL*1000; // scale factor
  lcd.setCursor(0, 1); // set cursor on LCD
  lcd.print("GAS_PPM:"); // print string on LCD
  lcd.setCursor(0, 2); // set cursor on LCD
  lcd.print(GAS_PPM); // print integer on LCD
  if (GAS_PPM >= 10000)  // compare gas levels
  {
   lcd.setCursor(0, 3); // set cursor on LCD
   lcd.print("Contents increase");  // print string on LCD
   digitalWrite(HOOTER_PIN, HIGH); // make D7 pin to HIGH
   delay(20); // wait for 20 mSec
  }
 else
  {
   lcd.setCursor(0, 3); // set cursor on LCD
   lcd.print("Contents Normal  "); // print string on LCD
   digitalWrite(HOOTER_PIN, LOW); // make D7 pin to LOW
   delay(20); // wait for 20 mSec
  }
}
```

Program to interface gas sensor with ADC0804 and 8051

```
#include<reg52.h>        //initializing of reg52 header file
#include<lcd.h>          //initializing of LCD header file
```

```
void disp_name();        //declaration of disp_name function
void delay();            //declaration of delay function
void disp_name1();       //declaration of disp_name1 function
char databyte[]={"0123456789"};
char a,b,c;
void main()      //definition of main function
{
P1=0xff;         //initialize the port 1 as 11111111
lcd_init();      //call the LCD initialize function
disp_name();     //call the disp_name function
lcd_init1();     //call the LCD1 initialize function
disp_name1();    //call the disp_name1 function

while(1)         //endless loop
{
i=P1; // read P1 port for getting analog sensor data
a=i%10; // add scaling factor
b=i/10; // add scaling factor
c=i/100; // add scaling factor
send_command(0xc0);       // send command to first location of LCD
send_data(databyte[c]);   //send first data
send_command(0xc1);       // send command to second location of LCD
send_data(databyte[b]);   //send second data
send_command(0xc2);       // send command to third location of LCD
send_data(databyte[a
]);          //send third data
delay();                  //delay for some moment
}
}

void disp_name()                //function to display data on first line of LCD
{
char i=0;
char databyte[]={"Gas Contents      "};       //Contents to be displayed
send_command(0x80);                           //Command of first line
while(databyte[i]!='\0')
{
send_data(databyte[i]);                       //send the data
i++;
}
}

void disp_name1()               //function to display data on second line of LCD
{
char i=0;
char databyte[]={"  Gas_ PPM "};   //contents to be displayed
send_command(0xC0);                //command of second line
while(databyte[i]!='\0')
{
send_data(databyte[i]);            //send the data
```

```
i++;
}
}

void delay()          //function to provide some delay
{
int i;
for(i=0;i<=250;i++);
}
```

9 Interfacing of 8051 and NuttyFi/ESP8266 with Digital Sensor

Electronic sensors which act on occurrence on an event and transmission takes place digitally are referred as digital sensors. Digital sensors provide output as logic "HIGH" or "LOW", which can be directly processed by the microcontroller and action can be taken accordingly. This chapter discusses the basic of digital sensors and its features. The interfacing of digital sensors with 8051 and NuttyFi is also discussed with the help of circuit diagram and programming.

9.1 PUSH BUTTON

A push button is an easy switch mechanism. Buttons usually made of plastic or metal. The surface can be flat or formed to accommodate the human finger on it. It is the simplest example of a digital sensor. Figure 9.1 shows the push buttons.

Figure 9.2 shows the block diagram for interfacing of push button with NuttyFi/8051 board. The system comprises of a +12V power supply, a +12V to +5V converter, a NuttyFi/8051, a LED with resistor 330 Ohm, and a button with breakout board. The objective is to read the status of button, if it is pressed then LED will be "ON" otherwise "OFF". Table 9.1 shows the list of the components required to design the system.

9.1.1 CIRCUIT DIAGRAM

Description of interfacing of push button with ESP8266/NuttyFi
The interfacing of the devices is as follows:

1. Connect output of +12V power supply to input of +12V to +5V convertor.
2. Connect output of +5V power supply to +5V of the NuttyFi/ESP8266.
3. Connect +Vcc, GND and OUT pin of the button to +5V, GND and D1 pins of the NuttyFi respectively.
4. Connect anode terminal of LED to D2 pin of NuttyFi.
5. Connect cathode terminal of LED to the ground.
6. Connect FTDI programmer to FTDI connector of the NuttyFi board.

Figure 9.3 shows the circuit diagram for interfacing of push button with NuttyFi.
Description of interfacing of push button with 8051
The interfacing of the devices is as follows:

1. Connect output of +12V power supply to input of +12V to +5V convertor.
2. Connect output of +5V power supply to Vcc of 8051.
3. Connect anode terminal of LED to P2.1 of 8051 through a 330 Ohm resistor.
4. Connect cathode terminal of LED to the ground.
5. Push to ON switch is connected to port P2.0 through a pull up resistance of 10k Ohm.

Figure 9.4 shows the circuit diagram for the interfacing of push button with 8051.

FIGURE 9.1   Push buttons.

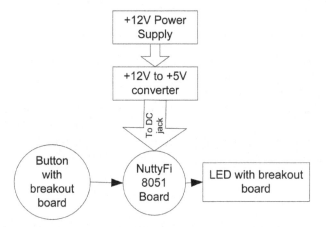

FIGURE 9.2   Block diagram for interfacing of push button with NuttyFi/8051 board.

TABLE 9.1

Components List to Design the System

Sr. No.	Name of the Components	Quantity	Specifications
	Components Used to Interface ESP8266 with Button		
1	+12V power supply	1	Output- +12V/1A
2	+12V to 5V convertor	1	Output- +5V/1A
3	NuttyFi	1	Analog pin-1
			Digital pin-10
			FTDI connector for programmer
4	Button with breakout board	1	Three pins
5	LED with breakout board	1	Red led
	Components Used to Interface 8051 with Button		
1	+12V power supply	01	Output- +12V/1A
2	+12V to 5V convertor	01	To regulate the input voltage to +5V/1A
3	330 Ohm resistor	01	0.25 Watt
4	LED	02	RED
5	8051 development board	01	Atmel Make including crystal oscillator and reset circuit
6	Button with breakout board	1	Three pins

9.1.2 Program

//Program to interface button sensor with ESP8266

```
const int BUTTON_Pin=D1;   // assign pin D1 to button
const int LED_PIN = D2; // assign pin D2 to LED pin
int BUTTON_Pin_STATE; // assume integer

void setup()
{
 pinMode(LED_PIN, OUTPUT);   // set D2 as an output
 pinMode(BUTTON_Pin, INPUT);  // set D1 as an input
}
void loop()
{
  BUTTON_Pin_STATE = digitalRead(BUTTON_Pin);// Read button pin
 if (BUTTON_Pin_STATE == HIGH) // check the state
 {
  digitalWrite(LED_PIN, HIGH); // make D2 pin to HIGH
  delay(20); // wait for 20mSec
 }
 else
 {
  digitalWrite(LED_PIN, LOW); // make D2 pin to LOW
  delay(20); // wait for 20mSec
 }
}
```

//Program to interface button sensor with 8051

```
#include <REG52.h>          // Initialization of header file
void delay();               // Declaration of delay function
void main()          //Definition of main function
{
P20=1;               // initially make the LED switch OFF
while(1)
{
if(P20==0)          //Check the output of the switch is 0 or not. (Switch is pressed or not)
{
P21=0;              //Glow the LED for some time
delay();            // Make the LED On for some time
P21=1;              //Switch OFF the LED
}
}
}
void delay()        //Delay function for some random delay generation
{
int i,j;
for(i=0;i<=100;i++)
{
for(j=0;j<=500;j++);
}
}
```

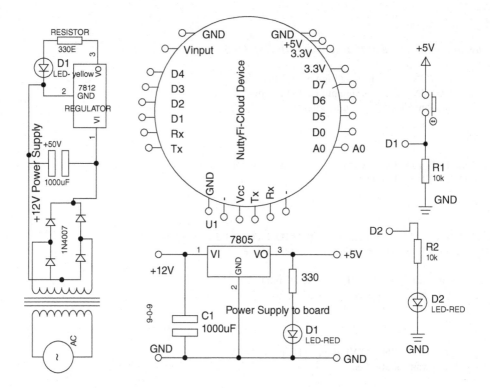

FIGURE 9.3 Circuit diagram for the interfacing of push button with NuttyFi/ESP8266.

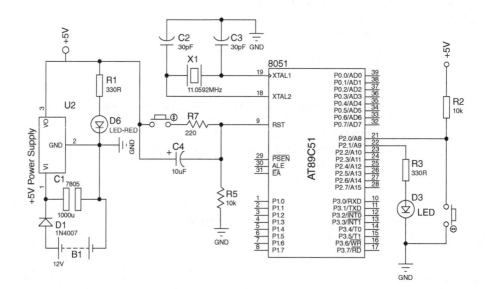

FIGURE 9.4 Circuit diagram for the interfacing of push button with 8051.

FIGURE 9.5 PIR sensor.

9.2 PIR SENSOR

PIR is a passive infrared device, which measures infrared (IR) light diverging from objects in its field of read. The objects with a temperature emit radiation, which is not visible to the human eye and falls in infrared wavelength range. The term passive refers to the fact that PIR devices don't generate or radiate energy for detection functions. They work entirely by infrared emission emitted by or mirrored from objects. Figure 9.5 shows the PIR sensor.

Figure 9.6 shows the block diagram for the interfacing of PIR sensor with NuttyFi/8051. The system comprises of a +12V power supply, a +12V to +5V converter, a NuttyFi/ESP8266, a LED, a resistor 330 Ohm, and a PIR sensor. The objective is to read the status of PIR sensor, if motion sensor state is true then LED will be "ON" otherwise "OFF". Table 9.2 shows the list of the components to design the system.

9.2.1 Circuit Diagram

Description of interfacing of a fire sensor with ESP8266

The interfacing of the devices is as follows:

1. Connect output of +12V power supply to input of +12V to +5V convertor.
2. Connect output of +5V power supply to +5V of the NuttyFi/ESP8266.

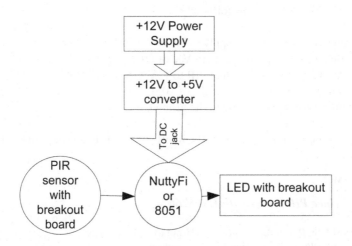

FIGURE 9.6 Block diagram of the system.

TABLE 9.2

Components List to Design the System

Sr. No.	Name of the Components	Quantity	Specifications
	Components Used to Interface ESP8266 with Fire Sensor		
1	+12V power supply	1	Output- +12V/1A
2	+12V to 5V convertor	1	Output- +5V/1A
3	NuttyFi	1	Analog pin-1
			Digital pin-10
			FTDI connector for programmer
4	LCD breakout board	1	4 Pins for +5V and 4 Pins for GND
5	LCD 20*4	1	16 pins LCD with backlight pins
6	PIR sensor with breakout board	1	Three pins
7	LED with breakout board	1	RED led
8	+12V power supply	1	Output- +12V/1A
9	+12V to 5V convertor	1	Output- +5V/1A
	Components Used to Interface 8051 with Fire Sensor		
1	+12V power supply	01	Output- +12V/1A
2	+12V to 5V convertor	01	To regulate the input voltage to +5V/1A
3	330 Ohm resistor	01	0.25 Watt
4	LED	02	RED
5	8051 development board	01	Atmel Make including crystal oscillator and reset circuit
6	PIR sensor with breakout board	1	Three pins

3. Connect +Vcc, GND and OUT pin of PIR to +5V, GND and D1 pins of the NuttyFi respectively.
4. Connect anode terminal of LED to D2 pin of NuttyFi.
5. Connect cathode terminal of LED to the ground.
6. Connect FTDI programmer to FTDI connector of the NuttyFi board.

Figure 9.7 shows the circuit diagram for the interfacing of PIR sensor with NuttyFi/ESP8266.

Description of interfacing of PIR sensor with 8051

The interfacing of the devices is as follows:

1. Connect output of +12V power supply to input of +12V to +5V convertor.
2. Connect output of +5V power supply to Vcc of 8051.
3. Connect +Vcc, GND and OUT pin of PIR to +5V, GND and P2.0 pins of 8051 respectively.
4. Connect anode terminal of LED to P2.1 of 8051.
5. Connect cathode terminal of LED to the ground.

Figure 9.8 shows the circuit diagram for the interfacing of PIR sensor with 8051.

9.2.2 PROGRAM

//Program to interface PIR sensor with ESP8266

```
const int MOTIONSENSOR_Pin=D1;    // assign D1 pin to motion sensor
const int INDICATOR_PIN = D2; // assign D2 pin to indicator
int MOTIONSENSOR_Pin_STATE; // assign integer
```

```
void setup()
{
 pinMode(INDICATOR_PIN, OUTPUT);   // set D2 as an output
 pinMode(MOTIONSENSOR_Pin, INPUT_PULLDOWN); // set D1 as an input
}

void loop()
{
  MOTIONSENSOR_Pin_STATE = digitalRead(MOTIONSENSOR_Pin);// Read PIR Sensor pin
if (MOTIONSENSOR_Pin_STATE == HIGH) // check state
 {
  digitalWrite(INDICATOR_PIN, HIGH); // set D2 pin to HIGH
  delay(20); // wait for 20 mSec
 }
 else
 {
  digitalWrite(INDICATOR_PIN, LOW); // set D2 pin to LOW
  delay(20); // wait for 20 mSec
 }
}
```

FIGURE 9.7   Circuit diagram for the interfacing of PIR sensor with NuttyFi/ESP8266.

//Program to interface PIR with 8051

```
#include <REG52.h>       // Initialization of header file
void delay();            // Declaration of delay function
void main()         //Definition of main function
{
P20=1;              // initially make the LED switch OFF
while(1)
{
if(P20==0)          //Check the output of the PIR sensor is it 0 or not.
{
P21=0;              //Glow the LED for some time
delay();            // Make the LED On for some time
P21=1;              //Switch OFF the LED
}
}
}


void delay()        //Delay function for some random delay generation
{
int i,j;
for(i=0;i<=100;i++)
{
for(j=0;j<=500;j++);
}
}
```

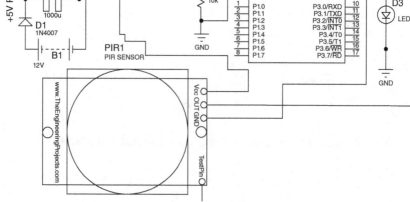

FIGURE 9.8   Circuit diagram for the interfacing of PIR sensor with 8051.

FIGURE 9.9 Flame sensor.

9.3 FLAME/FIRE SENSOR

A flame sensor is designed to detect the presence of a fire in the surroundings. It can be used in the applications like industrial furnaces, home safety, etc. A flame sensor responds or will respond quicker and accurate than a smoke or heat detector. Figure 9.9 shows fire sensor.

Figure 9.10 shows the block diagram for interfacing the fire sensor with NuttyFi/8051. The system comprises of a +12V power supply, a +12V to +5V converter, a NuttyFi/8051, a LED with resistor 330 Ohm, and a fire sensor. The objective is to read the status of fire sensor, if fire sensor state is true then LED will be "ON" otherwise "OFF". Table 9.3 shows the list of the components to design the system.

9.3.1 CIRCUIT DIAGRAM

Description of interfacing of fire sensor with ESP8266/NuttyFi

The interfacing of the devices is as follows:

1. Connect output of +12V power supply to input of +12V to +5V convertor.
2. Connect output of +5V power supply to +5V of the NuttyFi/ESP8266.
3. Connect +Vcc, GND and OUT pin of flame sensor to +5V, GND and D1 pins of the NuttyFi respectively.
4. Connect anode terminal of LED to D2 pin of NuttyFi.
5. Connect cathode terminal of LED to the ground.
6. Connect FTDI programmer to FTDI connector of the NuttyFi.

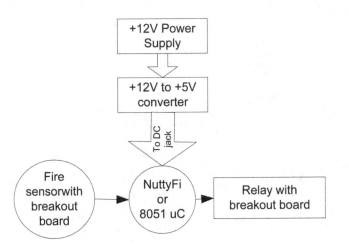

FIGURE 9.10 Block diagram of the system.

TABLE 9.3

Components List to Design the System

Sr. No.	Name of the Components	Quantity	Specifications
Components Used to Interface ESP8266 with Fire Sensor			
1	+12V power supply	1	Output- +12V/1A
2	+12V to 5V convertor	1	Output- +5V/1A
3	NuttyFi	1	Analog pin-1
			Digital pin-10
			FTDI connector for programmer
	LCD breakout board	1	4 Pins for +5V and 4 Pins for GND
	LCD 20*4	1	16 pins LCD with backlight pins
4	Fire sensor with breakout board	1	Three pins
5	Relay with breakout board	1	To run the hooter
Components Used to Interface 8051 with Fire Sensor			
1	+12V power supply	01	Output- +12V/1A
2	+12V to 5V convertor	01	To regulate the input voltage to +5V/1A
3	330 Ohm resistor	01	0.25 Watt
4	LED	02	RED
5	8051 development board	01	Atmel Make including crystal oscillator and reset circuit
6	Fire sensor with breakout board	1	Three pins

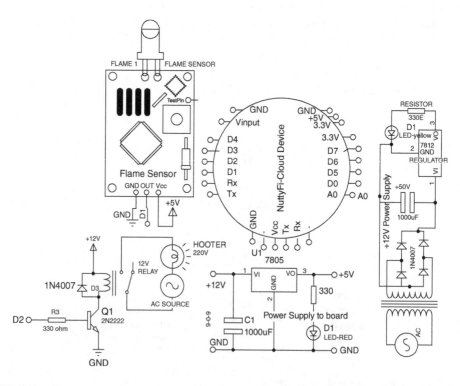

FIGURE 9.11   Circuit diagram for the interfacing of fire sensor with NuttyFi.

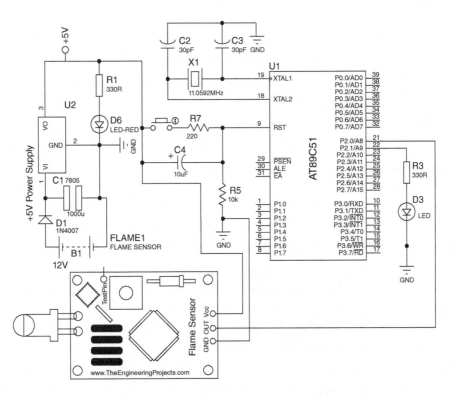

FIGURE 9.12   Circuit diagram for the interfacing of fire sensor with 8051.

Figure 9.11 shows the circuit diagram for interfacing fire sensor with NuttyFi.

Description of interfacing of fire sensor with 8051
The interfacing of the devices is as follows:

1. Connect output of +12V power supply to input of +12V to +5V convertor.
2. Connect output of +5V power supply to Vcc pin of 8051.
3. Connect +Vcc, GND and OUT pin of flame sensor to +5V, GND and P2.0 pins of 8051 respectively.
4. Connect anode terminal of LED to P2.1 pin of 8051.
5. Connect cathode terminal of LED to the ground.

Figure 9.12 shows the circuit diagram for interfacing fire sensor with 8051.

9.3.2   PROGRAM

//Program to interface fire sensor with ESP8266

```
const int FIRESENSOR_Pin=D1;   // assign pin D1 to fire sensor
const int RELAY_PIN = D2; // // assign pin D1 to relay
int FIRESENSOR_Pin_STATE; // assign integer
void setup()
{
  pinMode(RELAY_PIN, OUTPUT);     // set D2 pin as an output
  pinMode(FIRESENSOR_Pin, INPUT_PULLUP); // set D1 pin as an input
}
```

```
void loop()
{
  FIRESENSOR_Pin_STATE = digitalRead(FIRESENSOR_Pin);// Read fire Sensor pin
 if (FIRESENSOR_Pin_STATE == LOW) // check status
  {
   digitalWrite(RELAY_PIN, HIGH); // set D2 pin to HIGH
   delay(20); // wait for 20 msec
  }
  else
  {
   digitalWrite(RELAY_PIN, LOW); // set D2 pin to HIGH
   delay(20); // wait for 20 msec
  }
}
```

//Program to interface fire sensor with 8051

```
#include <REG52.h>          // Initialization of header file
void delay();               // Declaration of delay function

void main()         //Definition of main function
{
P20=1;              // initially make the LED switch OFF
while(1)
{
if(P20==0)          //Check the output of the fire sensor is it 0 or not.
{
P21=0;              //Glow the LED for some time
delay();            // Make the LED On for some time
P21=1;              //Switch OFF the LED
}
}
}

void delay()        //Delay function for some random delay generation
{
int i,j;
for(i=0;i<=100;i++)
{
for(j=0;j<=500;j++);
}
}
```

Section C

Interfacing of 8051 Microcontroller and NuttyFi/ESP8266 with Special Devices

10 Interfacing of 8051 and NuttyFi/ESP8266 with UART Based Devices

A universal asynchronous receiver-transmitter is used for asynchronous serial communication. In this the data format and transmission speeds are configurable. A device, the universal synchronous and asynchronous receiver-transmitter (USART), also supports synchronous operation. The universal asynchronous receiver-transmitter (UART) takes bytes of data and communicates the bits in a sequential fashion.

This chapter discusses the interfacing of devices in UART with 8051 and NuttyFi.

10.1 ULTRASONIC SENSOR

Ultrasonic module works on the principle of SONAR. The module has specification of working on UART. When a pulse of 10 μsec or more is given to the Trig pin, 8 pulses of 40 kHz are generated. Figure 10.1 shows the block diagram for interfacing of ultrasonic sensor (UART) with 8051. The system comprises of +12V power supply, +12V to +5V converter, 8051 microcontroller, 16x2 LCD, and UART. The objective of the system is to read the status of UART using 8051 and display on LCD. Table 10.1 shows the list of the components to design the system.

10.1.1 CIRCUIT DIAGRAM

Description of interfacing of ultrasonic sensor with 8051 in UART

1. Connect +12V power supply output to input of +12V to +5V convertor.
2. Connect +5V output to power up the 8051 board.
3. LCD is connected with port three in 4 pin mode.
4. Rx of UART is connected to port P3.0.
5. Connect one pin of buzzer to P2.3.
6. Connect another pin of buzzer to ground.
7. For programming user can use hardware programmer to program the 8051 uC or one can also use in system programming.

Figure 10.2 shows the connection diagram of the system.

10.1.2 PROGRAM

```
#include <REGX51.H>
#include <studio.h>
#include "lcd.h"
#include "utils.h"
sbit BUZZER=P2^3;
charsbuf[10], c, length;
charbuf[16];
unsigned char pos;
intrange;

charmygetchar(void)
{
```

```c
charc;
while(!RI);
RI =0;
c = SBUF;
myputchar(c); // echo to terminal
returnSBUF;
}
voidconvert()
{
range= (sbuf[0]-0x30)*100;
range= range+ (sbuf[1]-0x30)*10;
range= range+ (sbuf[2]-0x30)*1;
}
voidmain()
{
// -=-=- Intialize variables -=-=-=
BUZZER = 0;
SCON = 0x52; // 8-bit UART mode
TMOD = 0x20; // timer 1 mode 2 auto reload
TH1= 0xfd; // 9600 8-n-1
TR1 = 1; // run timer1
lcdInit();
lcdGotoXY(0,0); // 1st Line of LCD
lcdPrint("WELCOME TO ");
lcdGotoXY(0,1); // 2nd Line of LCD
lcdPrint("COLLISION SYSTEM");
delayms(5000); // 5 sec
pos= 0;
length= 0;
// -=-=- Program Loop -=-=-=
while(1)
{
c = mygetchar(); //loop till character received
if(c==0x0D) // if received character is <CR> end of line, time to display
{
length= pos;
pos= 0;
convert(); // convert serial buffer to integer
lcdClear();
lcdGotoXY(0,0); // 1st Line of LCD
sprintf (buf, "Range: %c%c%c%c%c%c cm", sbuf[0], sbuf[1], sbuf[2],
sbuf[3], sbuf[4], sbuf[5]);
lcdPrint(buf);
if(range <20) // we check if range is less than 20
{
BUZZER= 1;
lcdGotoXY(0,1); // 2nd Line of LCD
lcdPrint("Collision!!!!");
} else
{
BUZZER= 0;
lcdGotoXY(0,1); // 2nd Line of LCD
lcdPrint("Preset:20");
}
} else
{
sbuf[pos] = c;
pos++;
}
}
}
```

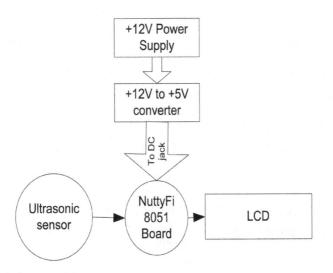

FIGURE 10.1   Block diagram of the system.

TABLE 10.1
Components List to Design the System

Sr. No.	Name of the Components	Quantity	Specifications
	Components Used to Interface 8051 uC with Ultrasonic Sensor		
1	+12V power supply	01	Output- +12V/1A
2	+12V to 5V convertor	01	To regulate the input voltage to +5V/1A
3	Crystal Oscillator	01	12 MHz
4	LCD	01	16x2
5	8051 development board	01	Atmel Make including crystal oscillator and reset circuit
6	Ultrasonic sensor	01	UART

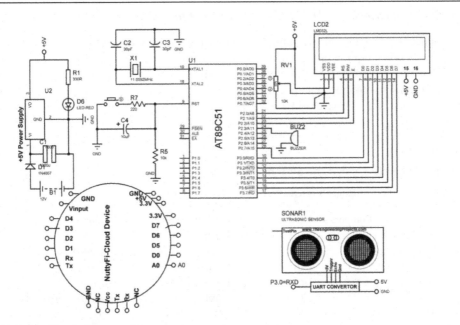

FIGURE 10.2   Circuit diagram of the system.

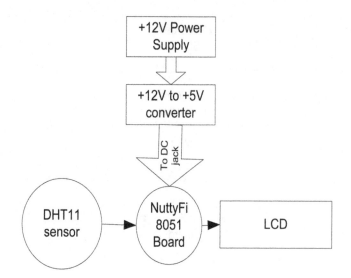

FIGURE 10.3   Block diagram of the system.

10.2   DHT11

DHT is very basic temperature and humidity sensor. The DHT sensors are made of two parts, a capacitive humidity sensor and a thermistor. The digital signal is fairly easy to read using any microcontroller.

Figure 10.3 shows the block diagram for interfacing of UART with 8051. The system comprises of +12V power supply, +12V to +5V converter, 8051 microcontroller, 16x2 LCD, and DHT11. The objective of the system is to read the DHT sensor using 8051 and display on LCD. Table 10.2 shows the list of the components to design the system.

10.2.1   Circuit Diagram

Description of interfacing of DHT11 sensor with 8051 in UART

1. Connect +12V power supply output to input of +12V to +5V convertor.
2. Connect +5V output to power up the 8051 board.
3. LCD is connected with port three in 4 pin mode.
4. Rx of UART is connected to port P1.0.
5. Connect one pin of buzzer to P2.3.

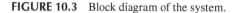

TABLE 10.2

Components List to Design the System

Sr. No.	Name of the Components	Quantity	Specifications
	Components Used to Interface 8051 uC with Ultrasonic Sensor		
1	+12V power supply	01	Output- +12V/1A
2	+12V to 5V convertor	01	To regulate the input voltage to +5V/1A
3	Crystal Oscillator	01	12 MHz
4	LCD	01	16x2
5	8051 development board	01	Atmel Make including crystal oscillator and reset circuit
6	DHT11 sensor	01	UART

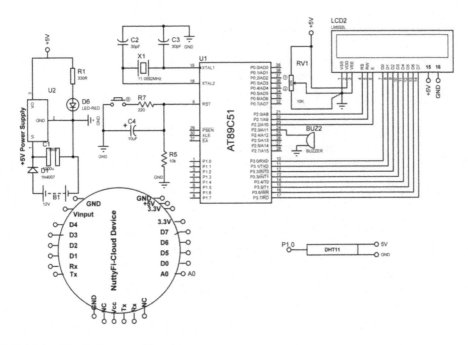

FIGURE 10.4 Circuit diagram of the system.

6. Connect another pin of buzzer to ground.
7. For programming user can use hardware programmer to program the 8051 uC or one can also use in system programming.

Figure 10.4 shows the connection diagram of the system.

10.2.2 PROGRAM

```
#include<reg51.h>
#include<stdio.h>
#include<string.h>
#include <stdlib.h>
#include "LCD16x2_4bit.h"

sbit DHT11=P2^1;                /* Connect DHT11 output Pin to P2.1 Pin */
intI_RH,D_RH,I_Temp,D_Temp,CheckSum;

void timer_delay20ms()          /* Timer0 delay function */
{
        TMOD = 0x01;
        TH0 = 0xB8;
        TL0 = 0x0C;
        TR0 = 1;
        while(TF0 == 0);
        TR0 = 0;
        TF0 = 0;

}

void timer_delay30us()          /* Timer0 delay function */
{
```

```
        TMOD = 0x01;                /* Timer0 mode1 (16-bit timer mode) */
        TH0 = 0xFF;             /* Load higher 8-bit in TH0 */
        TL0 = 0xF1;             /* Load lower 8-bit in TL0 */
        TR0 = 1;                /* Start timer0 */
        while(TF0 == 0);        /* Wait until timer0 flag set */
        TR0 = 0;                /* Stop timer0 */
        TF0 = 0;                /* Clear timer0 flag */
}

void Request()                  /* Microcontroller send request */
{
        DHT11 = 0;              /* set to low pin */
        timer_delay20ms();      /* wait for 20ms */
        DHT11 = 1;              /* set to high pin */
}

void Response()                         /* Receive response from DHT11 */
{
        while(DHT11==1);
        while(DHT11==0);
        while(DHT11==1);
}

intReceive_data()               /* Receive data */
{
        intq,c=0;
        for (q=0; q<8; q++)
        {
                while(DHT11==0);/* check received bit 0 or 1 */
                timer_delay30us();
                if(DHT11 == 1)          /* If high pulse is greater than 30ms */
                c = (c<<1)|(0x01);/* Then its logic HIGH */
                else                    /* otherwise its logic LOW */
                c = (c<<1);
                while(DHT11==1);
        }
        return c;
}

void main()
{
        unsigned char dat[20];
        LCD_Init();             /* initialize LCD */

        while(1)
        {
                Request();
                Response();

                I_RH=Receive_data();
                D_RH=Receive_data();
                I_Temp=Receive_data();
                D_Temp=Receive_data();
                CheckSum=Receive_data();

                if ((I_RH + D_RH + I_Temp + D_Temp) != CheckSum)
```

```
        {
                LCD_String_xy(0,0,"Error");
        }

        else
        {
                sprintf(dat,"Hum = %d.%d",I_RH,D_RH);
                LCD_String_xy(0,0,dat);
                sprintf(dat,"Tem = %d.%d",I_Temp,D_Temp);
                LCD_String_xy(1,0,dat);
                LCD_Char(0xDF);
                LCD_String("C");
                memset(dat,0,20);
                sprintf(dat,"%d   ",CheckSum);
                LCD_String_xy(1,13,dat);
        }
        delay(100);
    }
}
```

11 Interfacing of 8051 with I2C Based Devices

This chapter discusses the interfacing of 8051 with peripheral devices in I2C mode. I2C stands for inter IC communication. The interfacing of devices with 8051 and NuttyFi is discussed with the help of circuit diagram and programming.

11.1 DHT11

DHT11 is a single wire digital humidity and temperature sensor, which provides humidity and temperature values serially. It can measure relative humidity in percentage (20–90% RH) and temperature in degree Celsius in the range of 0–50°C. It has four pins of which two pins are connected with power supply, one is not used and the last one is used for data. The data is the only pin used for communication. Pulses of different TON and TOFF are decoded as logic "1" or logic "0" or start pulse or end of the frame.

Figure 11.1 shows the block diagram for interfacing of DHT11 with 8051. The system comprises of +12V power supply, +12V to +5V converter, 8051 microcontroller, 16x2 LCD, DHT11/22 sensor with breakout board, and Buzzer. The objective of the system is to read the status of DHT11/22 sensor using 8051. Table 11.1 shows the list of the components to design the system.

11.1.1 CIRCUIT DIAGRAM

Description of interfacing of DHT11 sensor with 8051

1. Connect +12V power supply output to input of +12V to +5V convertor.
2. Connect +5V output to power up the 8051 board.
3. LCD is connected with port three in 4-pin mode.
4. DHT11 sensor is connected to port P2.0.
5. For programming user can use hardware programmer to program the 8051 uC or one can also use in system programming also.

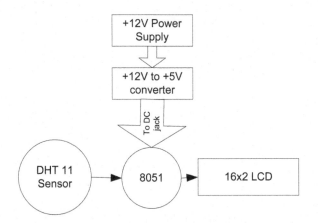

FIGURE 11.1 Block diagram for interfacing of DHT11 with 8051.

TABLE 11.1

Components List to Design the System

Sr. No.	Name of the Components	Quantity	Specifications
	Components Used to Interface 8051 uC with Fire Sensor		
1	+12V power supply	01	Output- +12V/1A
2	+12V to 5V convertor	01	To regulate the input voltage to +5V/1A
3	Crystal Oscillator	01	12 MHz
4	LCD	02	16x2
5	8051 development board	01	Atmel Make including crystal oscillator and reset circuit
6	DHT11/22 sensor	1	Three pins

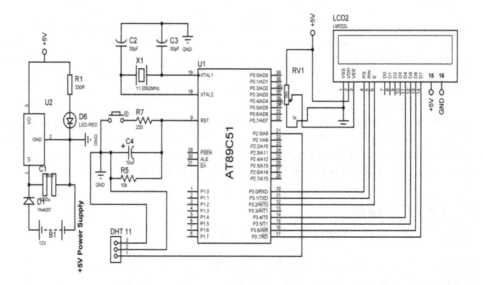

FIGURE 11.2 Circuit diagram for interfacing of DHT11 with 8051.

Figure 11.2 shows the circuit diagram of the system.

11.1.2 PROGRAM

Program to interface DHT11 sensor with 8051

```
#include<reg51.h>
#include "lcd.h"
#include<stdio.h>
#include<string.h>
#include <stdlib.h>

sbit DHT11=P2^1;               /* Connect DHT11 output Pin to P2.1 Pin */
int I_RH,D_RH,I_Temp,D_Temp,CheckSum;

void timer_delay20ms()          /* Timer0 delay function */
{
      TMOD = 0x01;
      TH0 = 0xB8;          /* Load higher 8-bit in TH0 */
      TL0 = 0x0C;          /* Load lower 8-bit in TL0 */
```

```
        TR0 = 1;                /* Start timer0 */
        while(TF0 == 0);        /* Wait until timer0 flag set */
        TR0 = 0;                /* Stop timer0 */
        TF0 = 0;                /* Clear timer0 flag */
}

void timer_delay30us()                    /* Timer0 delay function */
{
        TMOD = 0x01;                      /* Timer0 mode1 (16-bit timer mode) */
        TH0 = 0xFF;             /* Load higher 8-bit in TH0 */
        TL0 = 0xF1;             /* Load lower 8-bit in TL0 */
        TR0 = 1;                /* Start timer0 */
        while(TF0 == 0);        /* Wait until timer0 flag set */
        TR0 = 0;                /* Stop timer0 */
        TF0 = 0;                /* Clear timer0 flag */
}

void Request()                  /* Microcontroller send request */
{
        DHT11 = 0;              /* set to low pin */
        timer_delay20ms();      /* wait for 20ms */
        DHT11 = 1;              /* set to high pin */
}
void Response()                         /* Receive response from DHT11 */
{
        while(DHT11==1);
        while(DHT11==0);
        while(DHT11==1);
}

int Receive_data()              /* Receive data */
{
        int q,c=0;
        for (q=0; q<8; q++)
        {
                while(DHT11==0);/* check received bit 0 or 1 */
                timer_delay30us();
                if(DHT11 == 1)          /* If high pulse is greater than 30ms */
                c = (c<<1)|(0x01);/* Then its logic HIGH */
                else                    /* otherwise its logic LOW */
                c = (c<<1);
                while(DHT11==1);
        }
        return c;
}

void main()
{
        unsigned char dat[20];
        LCD_Init();             /* initialize LCD */

        while(1)
        {
                Request();      /* send start pulse */
                Response();     /* receive response */
```

```
I_RH=Receive_data();          /* store first eight bit in I_RH */
D_RH=Receive_data();          /* store next eight bit in D_RH */
I_Temp=Receive_data();        /* store next eight bit in I_Temp */
D_Temp=Receive_data();        /* store next eight bit in D_Temp */
CheckSum=Receive_data();/* store next eight bit in CheckSum */
if ((I_RH + D_RH + I_Temp + D_Temp) != CheckSum)
{
        LCD_String_xy(0,0,"Error");
}

else
{
        sprintf(dat,"Hum = %d.%d",I_RH,D_RH);
        LCD_String_xy(0,0,dat);
        sprintf(dat,"Tem = %d.%d",I_Temp,D_Temp);
        LCD_String_xy(1,0,dat);
        LCD_Char(0xDF);
        LCD_String("C");
        memset(dat,0,20);
        sprintf(dat,"%d   ",CheckSum);
        LCD_String_xy(1,13,dat);
}

delay(100);
}
}
```

11.2   ULTRASONIC SENSOR

Ultrasonic Module HC-SR04 works on the principle of SONAR system. The HC-SR04 module has ultrasonic transmitter, receiver, and control circuit on a single board.

The module has four pins, Vcc, Gnd, Trig, and Echo. When a pulse of 10μsec or more is given to the Trig pin, eight pulses of 40 kHz are generated. After this, the Echo pin is made high by the control circuit in the module. Echo pin remains high till it gets echo signal of the transmitted pulses back. The time for which the echo pin remains high, that is, the width of the Echo pin gives the time taken for generated ultrasonic sound to travel toward the object and return. Using this time and the speed of sound in air, we can find the distance of the object by a simple formula for distance using speed and time.

Figure 11.3 shows the block diagram for interfacing of ultrasonic sensor with 8051. The system comprises of +12V power supply, +12V to +5V converter, 8051 microcontroller, 16x2 LCD, and Ultrasonic sensor with breakout board. The objective of the system is to read the status of Ultrasonic sensor using 8051. Table 11.2 shows the list of the components to design the system.

11.2.1   CIRCUIT DIAGRAM

Description of interfacing of ultrasonic sensor with 8051

1. Connect +12V power supply output to input of +12V to +5V convertor.
2. Connect +5V output to power up the 8051 board.
3. LCD is connected with port three in 4-pin mode.
4. Ultrasonic sensor is connected to port P2.0 and P2.1.
5. For programming user can use hardware programmer to program the 8051 uC or one can also use in system programming.

Figure 11.4 shows the connection diagram of the system.

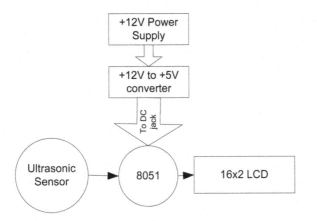

FIGURE 11.3 Block diagram for interfacing of ultrasonic sensor with 8051.

TABLE 11.2
Components List to Design the System

Sr. No.	Name of the Components	Quantity	Specifications
	Components Used to Interface 8051 uC with Ultrasonic Sensor		
1	+12V power supply	01	Output- +12V/1A
2	+12V to 5V convertor	01	To regulate the input voltage to +5V/1A
3	Crystal Oscillator	01	12 MHz
4	LCD	01	16x2
5	8051 development board	01	Atmel Make including crystal oscillator and reset circuit
6	Ultrasonic sensor	1	Four pins

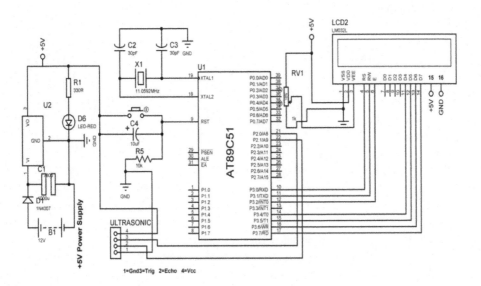

FIGURE 11.4 Circuit diagram for interfacing of ultrasonic sensor with 8051.

11.2.2 PROGRAM

```c
#include<reg52.h>
#include <stdio.h>
#include <lcd.h>
#include <math.h>
#define sound_velocity 34300           /* sound velocity in cm per second */
#define period_in_us pow(10,-6)
#define Clock_period 1.085*period_in_us              /* period for clock cycle of 8051*/

sbit Trigger_pin=P2^0;            /* Trigger pin */
sbit Echo_pin=P2^1;              /* Echo pin */

void main()
{
        float distance_measurement, value;
        unsigned char distance_in_cm[10];
        lcd_init();                    /* Initialize 16x2 LCD */
        display(0x80, "Distance:");
        init_timer();                      /* Initialize Timer*/

        while(1)
        {
                send_trigger_pulse();          /* send trigger pulse of 10us */
                while(!Echo_pin);              /* Waiting for Echo */
                TR0 = 1;                       /* Timer Starts */
                while(Echo_pin && !TF0);       /* Waiting for Echo goes LOW */
                TR0 = 0;                       /* Stop the timer */

                /* calculate distance using timer */
                value = Clock_period * sound_velocity;
                distance_measurement = (TL0|(TH0<<8));       /* read timer register for time count */
                distance_measurement = (distance_measurement*value)/2.0; /* find distance(in cm) */
                display(0xC0, distance_in_cm);         /* show distance on 16x2 LCD */
                display (0x80, " cm ");
                delay(100);
        }
}

void Delay_us()
{
        TL0=0xF5;
        TH0=0xFF;
        TR0=1;
        while (TF0==0);
        TR0=0;
        TF0=0;
}

void init_timer()
{
        TMOD=0x01;                 /*initialize Timer*/
        TF0=0;
        TR0 = 0;
}

void send_trigger_pulse()
```

```
{
        Trigger_pin= 1;              /* pull trigger pin HIGH */
        Delay_us();                  /* provide 10uS Delay*/
        Trigger_pin = 0;             /* pull trigger pin LOW*/
}

void display(char x, char databyte[])
{
        char y=0;
        send_command(x);
        while(databyte[y]!='\0')
        {
                send_data(databyte[y]);
                y++;
        }
}
```

11.3   EEPROM

An EEPROM (Electrically Erasable Programmable Read Only Memory) is a non-volatile flash memory, which has the capability to retain data even if the power is removed. The most commonly used EEPROM family is 24CXX series such as 24C02, 24C04, 24C08Â, etc. The most common IC is EEPROMIC24C04. Interfacing an EEPROM to microcontroller is simple. Only need to make two connections between the 24C04 IC and 8051 microcontroller.

Figure 11.5 shows the block diagram for interfacing EEPROM with 8051. The system comprises of +12V power supply, +12V to +5V converter, 8051 microcontroller, 16x2 LCD, and EEPROM with breakout board. The objective of the system is to store and read the data using 8051. Table 11.3 shows the list of the components to design the system.

11.3.1   CIRCUIT DIAGRAM

Description of interfacing of EEPROM with 8051

1. Connect +12V power supply output to input of +12V to +5V convertor.
2. Connect +5V output to power up the 8051 board.

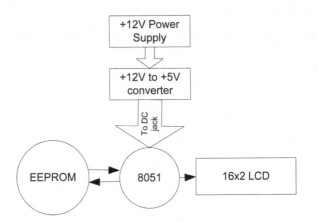

FIGURE 11.5   Block diagram for interfacing EEPROM with 8051.

TABLE 11.3

Components List to Design the System

Sr. No.	Name of the Components	Quantity	Specifications
	Components Used to Interface 8051 uC with Ultrasonic Sensor		
1	+12V power supply	01	Output- +12V/1A
2	+12V to 5V convertor	01	To regulate the input voltage to +5V/1A
3	Crystal Oscillator	01	12 MHz
4	LCD	02	16x2
5	8051 development board	01	Atmel Make including crystal oscillator and reset circuit
6	EEPROM	1	24C04

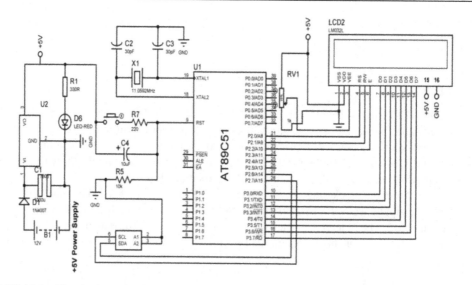

FIGURE 11.6　Circuit diagram for interfacing EEPROM with 8051.

3. LCD is connected with port three in 4-pin mode.
4. EEPROM is connected to port P2.6 and P2.7.
5. For programming user can use hardware programmer to program the 8051 uC or one can also use in system programming.

Figure 11.6 shows the circuit diagram for interfacing EEPROM with 8051.

11.3.2 PROGRAM

```
#include<reg52.h>
sbit REG_MSB=B^7;
sbit REG_LSB=B^0;
sbit SDA=P2^6;
sbit SCL=P2^7;
void delay();

void start();
void stop();
void send_data(char);
char receive_data();
```

```
void main()
{
        start();
        send_data(0xa0);                        // device address
        send_data(0x01);                        // word address
        send_data(0xaa);                        // data send
        stop();
        start();
        send_data(0xa1);                        // device address
        receive_data();
        stop();

        P0=B;
}
/////////////////////////////////////////////
void send_data(char x)
{
        int i;
        B=x;
        for(i=0;i<8;i++)
        {
                SCL=0;
                delay();
                SDA=REG_MSB;
                SCL=1;
                delay();
                SCL=0;
                delay();
                B=B<<1;
        }
        SDA=1;
        SCL=1;
        delay();
        SCL=0;
        delay();
}
/////////////////////////////////////////////
char receive_data()
{
        int i;
        for(i=0;i<8;i++)
        {
                SCL=1;
                delay();
                SCL=0;
                delay();
                REG_LSB=SDA;
                B=B<<1;
                delay();
        }
        return B;
}
/////////////////////////////////////////////
void start()
{
        SCL=0;
```

```
        SDA=1;
        delay();
        SCL=1;
        delay();
        SDA=0;
        delay();
        SCL=0;
}
/////////////////////////////////////////
void stop()
{
        delay();
        SDA=0;
        SCL=1;
        delay();
        SDA=1;
        delay();
        SCL=0;
}
/////////////////////////////////////////
void delay()
{
int i;
for(i=0;i<=100;i++);
}
```

11.4   DS1307

A Real Time Clock module is basically a time tracking device, which gives the current time and date. The most commonly used RTC module is DS1307 IC. Only two connections are required to connect between the RTC module and 8051.

Figure 11.7 shows the block diagram for interfacing RTC with 8051. The system comprises of +12V power supply, +12V to +5V converter, 8051 microcontroller, 16x2 LCD, and EEPROM with breakout board. The objective of the system is to store and read the data using 8051.

Table 11.4 shows the list of the components to design the system.

11.4.1   Circuit Diagram

Description of interfacing of DS1307 with 8051
    Figure 11.8 shows the circuit diagram for interfacing RTC with 8051.

1. Connect +12V power supply output to input of +12V to +5V convertor.
2. Connect +5V output to power up the 8051 board.
3. LCD is connected with port three in 4-pin mode.
4. DS1307 is connected to port P2.6 and P2.5.
5. For programming user can use hardware programmer to program the 8051 uC or one can also use in system programming.

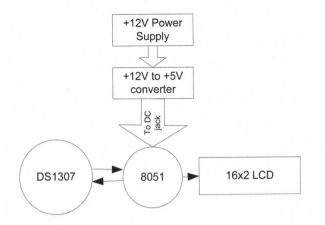

FIGURE 11.7   Block diagram for interfacing RTC with 8051.

TABLE 11.4
Components List to Design System

Sr. No.	Name of the Components	Quantity	Specifications
	Components Used to Interface 8051 uC with Ultrasonic Sensor		
1	+12V power supply	01	Output- +12V/1A
2	+12V to 5V convertor	01	To regulate the input voltage to +5V/1A
3	Crystal Oscillator	01	12 MHz
4	LCD	02	16x2
5	8051 development board	01	Atmel Make including crystal oscillator and reset circuit
6	RTC	1	DS1307

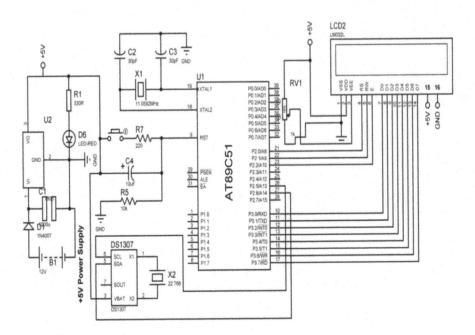

FIGURE 11.8   Circuit diagram for interfacing RTC with 8051.

11.4.2 PROGRAM

```
#include<reg51.h>
sbit SCL=P2^5;
sbit SDA=P2^6;
void start();
void write(unsigned char);
delay(unsigned char);

void main()
{
start();
write(0xA2); //slave address//
write(0x00); //control register address//
write(0x00); //control register 1 value//
write(0x00); //control regiter2 value//
write (0x28); //sec value//
write(0x50) ;//minute value//
write(0x02);//hours value//
}

void start()
{
SDA=1; //processing the data//
SCL=1; //clock is high//
delay(100);
SDA=0; //sent the data//
delay(100);
SCL=0; //clock signal is low//
}
void write(unsigned char d)
{
unsigned char k, j=0x80;
for(k=0;k<8;k++)
{
SDA=(d&j);
J=j>>1;
SCL=1;
delay(4);
SCL=0;
}
SDA=1;
SCL=1;
delay(2);
c=SDA;
delay(2);
SCL=0;
}

void delay(int p)
{
unsigned int a,b;
For(a=0;a<255;a++); //delay function//
For(b=0;b<p;b++);
}
```
//Read Operation from Slave to Master

```
#include<reg51.h>
sbit SCL=P2^5;
sbit SDA=P2^6;
void start();
void write(usigned char );
void read();
void ack();
void delay(unsigned char);

void main()
{
start();
write(0xA3);// slave address in read mode//
read();
ack();
sec=value;
}
void start()
{
SDA=1; //processing the data//
SCL=1; //clock is high//
delay(100);
SDA=0; //sent the data//
delay(100);
SCL=0; //clock signal is low//
}
void write(unsigned char d)
{
unsigned char k, j=0×80;
for(k=0;k<8;k++)
{
SDA=(d&j);
J=j>>1;
SCL=1;
delay(4);
SCL=0;
}
SDA=1;
SCL=1;
delay(2);
c=SDA;
delay(2);
SCL=0;
}
void delay(int p)
{
unsignedinta,b;
For(a=0;a<255;a++); //delay function//
For(b=0;b<p;b++);
}
Void read ()
{
Unsigned char j, z=0×00, q=0×80;
SDA=1;
for(j=0;j<8;j++)
{
```

```
SCL=1;
delay(100);
flag=SDA;
if(flag==1)
{
z=(z|q);
q=q>>1;
delay (100);
SCL=0;
}
void ack()
{
SDA=0; //SDA line goes to low//
SCL=1; //clock is high to low//
delay(100);
SCL=0;
}
```

12 Interfacing of 8051 with SPI Based Devices

The SPI communication stands for serial peripheral interface communication protocol, which was developed by the Motorola in 1972. SPI interface can be done on popular communication controllers such as 8051, PIC, AVR, ARM controller, etc. It has synchronous serial communication data link that operates in full duplex, which means the data signals carry on both the directions simultaneously.

12.1 INTRODUCTION

SPI protocol consists of four wires such as MISO, MOSI, CLK, and SS used for master/slave communication. The master is a microcontroller, and the slaves are other peripherals like sensors, GSM modem, GPS modem, etc. The multiple slaves are interfaced to the master through a SPI serial bus. The SPI protocol does not support the Multi-master communication and it is used for a short distance within a circuit board.

12.1.1 SPI LINES

MISO (Master in Slave Out): The MISO line is configured as an input in a master device and as an output in a slave device.

MOSI (Master out Slave In): The MOSI is a line configured as an output in a master device and as an input in a slave device wherein it is used to synchronize the data movement.

SCK (Serial Clock): This signal is always driven by the master for synchronous data transfer between the master and the slave. It is used to synchronize the data movement both in and out through the MOSI and MISO lines.

SS (Slave Select) and CS (Chip Select): This signal is driven by the master to select individual slaves/peripheral devices. It is an input line used to select the slave devices.

Figure 12.1 shows the block diagram to interface SPI device with 8051. The system comprises of +12V power supply, +12V to +5V converter, 8051 microcontroller, 16x2 LCD, and EEPROM with breakout board. The objective of the system is to interface the devices in SPI mode with 8051. Table 12.1 shows the list of the components to design the system.

12.2 CIRCUIT DIAGRAM

Description of interfacing of SPI device with 8051

1. Connect +12V power supply output to input of +12V to +5V convertor.
2. Connect +5V output to power up the 8051 board.
3. LCD is connected with port three in 4-pin modes.
4. SPI sensor is connected to P1.4 to P1.7.
5. For programming user can use hardware programmer to program the 8051 uC or one can also use in system programming.

Figure 12.2 shows the circuit diagram of the system.

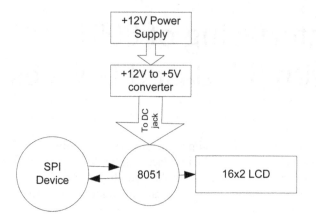

FIGURE 12.1   Block diagram of the system.

TABLE 12.1

List of the Components to Design the System

Sr. No.	Name of the Components	Quantity	Specifications
1	+12V power supply	01	Output- +12V/1A
2	+12V to 5V convertor	01	To regulate the input voltage to +5V/1A
3	Crystal Oscillator	01	12 MHz
4	LCD	02	16x2
5	8051 development board	01	Atmel Make including crystal oscillator and reset circuit
6	SPI enabled device	1	Any SPI device

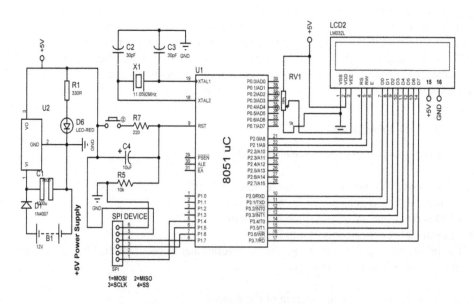

FIGURE 12.2   Circuit diagram of the system.

12.3   PROGRAM

```
#define char unsigned char

sbit RF_MISO= P2^3;  //say
sbit RF_MOSI= P2^2;
sbit RF_SCK= P2^1;
sbit RF_CSN= P2^0;

void spiWr(uchar);
uchar spiRd(void);

/*Low level SPI WRITE FUNCTION*/
void spiWr(uchar dat)
{
/* software SPI, send MSB first */
static uchar i,c;
c = dat;
for(i=0;i<8;i++)
{
if((c&0x80)==0x80)
RF_MOSI = 1;
else
RF_MOSI = 0;
RF_SCK = 1;
c=c<<1;
RF_SCK = 0;
}
}

/* Low level SPI READ FUNCTION*/
char spiRd(void)
{
/* software SPI read, MSB read first */
static uchar i, dat;
for(i=0;i<8;i++)
{
dat = dat<<1;
RF_SCK = 1;
if(RF_MISO)
dat = dat+1;
RF_SCK = 0;
}
return dat;
}
```

13 Interfacing of 8051 with One Wire Interface-based Devices

This chapter explores one wire hardware and software interfacing with 8051 microcontroller. The interfacing of devices with 8051 and NuttyFi is discussed with the help of circuit diagram and programming.

13.1 DS1820

DS18B20 is a temperature sensor, which can measure temperature from −55°C to +125°C with an accuracy of ±5%. It can be used in industrial applications where high accuracy is required. Data received from a single wire is of 9–12 bit. Sensor has three pins in total, Vcc, Gnd, and data pin.

Figure 13.1 shows the block diagram for the system. The system comprises +12V power supply, +12V to +5V converter, 8051 microcontroller, 16x2 LCD, and DS1820 sensor with breakout board. The objective of the system is to read the data from sensor with one wire protocol. Table 13.1 shows the list of the components to design the system.

13.1.1 Circuit Diagram

Description of interfacing of DS1820 with 8051

1. Connect +12V power supply output to input of +12V to +5V convertor.
2. Connect +5V output to power up the 8051 board.
3. LCD is connected with port three in 4-pin mode.
4. DS1820 is connected to port P1.7.
5. For programming user can use hardware programmer to program the 8051 uC or one can also use in system programming.

Figure 13.2 shows the circuit diagram of the system.

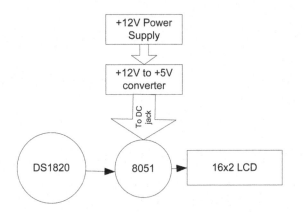

FIGURE 13.1   Block diagram of the system.

TABLE 13.1

List of the Components to Design the System

Sr. No.	Name of the Components	Quantity	Specifications
1	+12V power supply	01	Output- +12V/1A
2	+12V to 5V convertor	01	To regulate the input voltage to +5V/1A
3	Crystal Oscillator	01	12 MHz
4	LCD	02	16x2
5	8051 development board	01	Atmel Make including crystal oscillator and reset circuit
6	Sensor	1	DS 1820

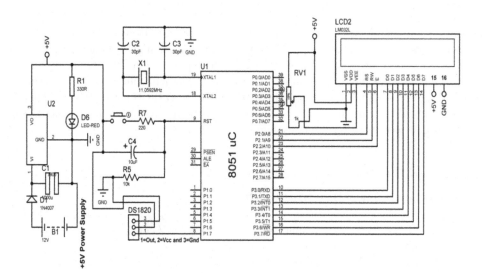

FIGURE 13.2   Circuit diagram of the system.

13.1.2   PROGRAM

include<reg51.h>

sbit dq = P1^7; // connect with DS1820 Data pin

sbit rs=P3^0;
sbit en=P3^1;

void delay_ms(int j)
{
unsigned char i;

for(;j;j--)
for(i=122;i<=0;i--);
}

void data_lcd(unsigned char dat)
{
P2=dat;

```
rs=1;
en=1;
delay_ms(200);
en=0;
}
void cmd_lcd(unsigned char cmd)
{
P2=cmd;
rs=0;
en=1;
delay_ms(200);
en=0;
}

void init_lcd()
{
cmd_lcd(0x38);
cmd_lcd(0x01);
cmd_lcd(0x0c);
}

void str_lcd(unsigned char *p)
{
while(*p)
data_lcd(*p++);
}
void delayus(int us)
{
int i;
for (i=0; i<us; i++);
}

bit reset(void)
{
bit presence;
dq = 0;
delayus(29);
dq = 1;
delayus(3);
presence = dq;
delayus(25);
return(presence);
}

bit readbit(void)
{
unsigned char i=0;
dq = 0;
dq=1;
for (i=0; i < 3; i++);
return(dq);
}

void writebit(bit Dbit)
{
unsigned char i=0;
```

```c
dq=0;
dq = Dbit?1:0;
delayus(5);
dq = 1;
}

unsigned char readbyte(void)
{
unsigned char i;
unsigned char din = 0;
for (i=0;i<8;i++)
{
din|=readbit()? 0x01<<i:din;
delayus(6);
}
return(din);
}
void writebyte(unsigned char dout)
{
unsigned char i;
for (i=0; i<8; i++)
{
writebit((bit)(dout & 0x1));
dout = dout >> 1;
}
delayus(5);
}

unsigned char * ReadTemp()
{
unsigned char n;
unsigned char buff[2]=0;
reset();

writebyte(0xcc);
writebyte(0x44);

while (readbyte()==0xff);
delay_ms(500);
reset();

writebyte(0xcc);
writebyte(0xbe);

for (n=0; n<9; n++)
buff[n]=readbyte();

return buff;
}

void int_lcd(int dat)
{
int str[5]={0},i=0;
if(dat==0)
data_lcd('0');
```

```
else
while(dat>0)
{
str[i]= (dat%10)+48;
dat=dat/10;
i++;
}
i--;
for(;i>=0;i--)
data_lcd(str[i]);
}

void main()
{
unsigned char tp,*temp,t=0x00;
init_lcd();
cmd_lcd(0x80);
while(1)
{

temp=ReadTemp();
temp[1]=temp[1]&0x07;
tp=temp[0]>>4;
temp[1]=temp[1]<<4;
tp=tp+temp[1];
cmd_lcd(0x80);
str_lcd("temperature is ");
cmd_lcd(0xc0);

int_lcd(tp);
data_lcd(223);
}
}
```

13.2   8051 TO 8051 COMMUNICATION

Figure 13.3 shows the block diagram to interface two 8051. The system comprises of +12V power supply, +12V to +5V converter, and two 8051 microcontroller board with breakout board. The objective of the system is to transfer the data from one 8051 microcontroller to another using Rx and Tx pins. Table 13.2 shows the list of the components to design the system.

13.2.1 CIRCUIT DIAGRAM

Description of interfacing of 8051 with 8051

The interfacing of 8051 with other 8051 is as per the given guidelines. Figure 13.4 shows the connection diagram of the system.

1. Connect +12V power supply output to input of +12V to +5V convertor.
2. Connect +5V output to power up the 8051 board.
3. LCD is connected with port three in 4-pin modes as per the above diagram.
4. Rx and Tx pins of one controller is connected to other Tx and Rx pins of controller.
5. For programming user can use hardware programmer to program the 8051 uC or one can also use in system programming.

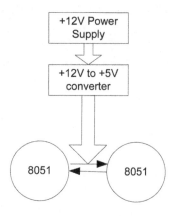

FIGURE 13.3   Block diagram to interface two 8051.

TABLE 13.2

List of the Components to Design the System

Sr. No.	Name of the Components	Quantity	Specifications
1	+12V power supply	01	Output- +12V/1A
2	+12V to 5V convertor	01	To regulate the input voltage to +5V/1A
3	Crystal Oscillator	01	12 MHz
4	Two 8051 development board	01	Atmel Make including crystal oscillator and reset circuit

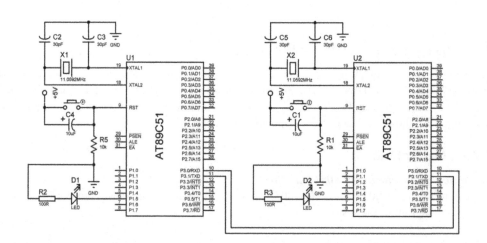

FIGURE 13.4   Circuit diagram of the system.

13.2.2 PROGRAM

Serial transmit program:

```c
#include<reg52.h>
#include<lcd.h>
void disp_key(char key,unsigned char);
void delay();
void transmit(char);
char key;
char add=0x80;
char databyte[]="ABCDEFGHIJKLMNOP";
/////////////////////////////////////////////////////////////
void main()
{
        TMOD=0x20;      // timer 1 8 bit auto reload
        TH1=0xfa;       // 4800 baud rate
        SCON=0x50;      // serial mode 1 8 bit data 1 start bit 1 stop bit.
        TR1=1;                  // start the timer
        TI=0;
        RI=0;
        lcd_init();
        while(1)
        {               P0=0xfe;
                        key=P0;
                        disp_key(key,add);

                        P0=0xfd;
                        key=P0;
                        disp_key(key,add);

                        P0=0xfb;
                        key=P0;
                        disp_key(key,add);

                        P0=0xf7;
                        key=P0;
                        disp_key(key,add);
        }
}
/////////////////////////////////////////////////////////////
void disp_key(char key,unsigned char add)
{
        switch(key)
        {
                send_command(add);
                case 0xee:                      // A
                        send_data(databyte[0]);         transmit('A'); delay(); add++;;
                        break;
                case 0xDe:                      // B
                        send_data(databyte[1]);         transmit('B'); delay(); add++;;
                        break;
                case 0xBe:                      // C
                        send_data(databyte[2]);         transmit('C'); delay(); add++;;
                        break;
```

```
        case 0x7e:                      // D
                         send_data(databyte[3]);        transmit('D'); delay(); add++;;
                         break;
        case 0xeD:                      // E
                         send_data(databyte[4]);         transmit('E');delay(); add++;;
                         break;
        case 0xDD:                      // F
                         send_data(databyte[5]);        transmit('F'); delay(); add++;;
                         break;
        case 0xBD:                      // G
                         send_data(databyte[6]);        transmit('G'); delay(); add++;;
                         break;
        case 0x7D:                      // H
                         send_data(databyte[7]);        transmit('H'); delay(); add++;;
                         break;
        case 0xeB:                      // I
                         send_data(databyte[8]);        transmit('I'); delay(); add++;;
                         break;
        case 0xDB:                      // J
                         send_data(databyte[9]);        transmit('J'); delay(); add++;;
                         break;
        case 0xBB:                      // K
                         send_data(databyte[10]); transmit('K');        delay(); add++;;
                         break;
        case 0x7B:                      // L
                         send_data(databyte[11]); transmit('L');        delay(); add++;;
                         break;
        case 0xE7:                      // M
                         send_data(databyte[12]); transmit('M');        delay(); add++;;
                         break;
        case 0xD7:                      // N
                         send_data(databyte[13]); transmit('N');        delay(); add++;;
                         break;
        case 0xB7:                      // O
                         send_data(databyte[14]); transmit('O');        delay(); add++;;
                         break;
        case 0x77:                      // P
                         send_data(databyte[15]); transmit('P');        delay(); add++;;
                         break;
        }
}
////////////////////////////////////////////////////////////
void transmit(char x)
{
        SBUF=x;                         // place value in buffer
        while(TI==0);           // wait till transmited
        TI=0;
}
////////////////////////////////////////////////////////////
void delay()
{
        int i,j;
        for(i=0;i<=75;i++)
        {
                for(j=0;j<=2500;j++)
                {}
```

```
        }
}
```

Serial receive:

```
#include<reg52.h>
void main()
{
        char byte;
        TMOD=0x20;      // timer 1 8 bit auto reload
        TH1=0xfa;       // 4800 baud rate
        SCON=0x50;      // serial mode 1 8 bit data 1 start bit 1 stop bit.
        TR1=1;                  // start the timer
        TI=0;
        RI=0;
        while(1)
        {
                while(RI==0);
                {
                        byte=SBUF;
                        P0=byte;
                        RI=0;
                }
        }
}
```

14 Interfacing of 8051 with Bluetooth

This chapter discusses the interfacing of 8051 with Bluetooth. The interfacing of devices with 8051 and NuttyFi is discussed with the help of circuit diagram and programming.

Bluetooth is a wireless technology standard used for exchanging data between fixed and mobile devices over short distances. It works on UHF radio waves in the industrial, scientific, and medical radio bands, from 2.402 GHz to 2.480 GHz.

14.1 INTRODUCTION

HC-05 is a Bluetooth device used for wireless communication. It works on serial communication (UART). It is a 6-pin module. It can be used in two modes; data mode and command mode. AT commands are required in command mode. The module works on 5V or 3.3V. As HC-05 Bluetooth module has 3.3 V level for Rx/Tx and microcontroller can detect 3.3 V level.

Figure 14.1 shows the block diagram for interfacing of Bluetooth with 8051. The system comprises of +12V power supply, +12V to +5V converter, 8051 microcontroller, 16x2 LCD, Bluetooth with breakout board, and Buzzer. The objective of the system is to make LED "ON" and "OFF" with Bluetooth. Table 14.1 shows the list of the components to design the system.

14.2 CIRCUIT DIAGRAM

Description of interfacing of Bluetooth with 8051

1. Connect +12V power supply output to input of +12V to +5V convertor.
2. Connect +5V output to power up the 8051 board.
3. Rx of Bluetooth is connected to port P3.1.
4. Tx of Bluetooth is connected to P3.0.

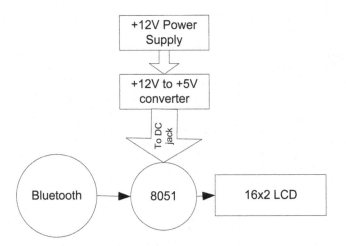

FIGURE 14.1 Block diagram for interfacing of Bluetooth with 8051.

TABLE 14.1

Components List to Design the System

Sr. No.	Name of the Components	Quantity	Specifications
	Components Used to Interface 8051 uC with Fire Sensor		
1	+12V power supply	01	Output- +12V/1A
2	+12V to 5V convertor	01	To regulate the input voltage to +5V/1A
3	Crystal Oscillator	01	12 MHz
4	LCD	02	16x2
5	8051 development board	01	Atmel Make including crystal oscillator and reset circuit
6	Bluetooth	1	Four pin

5. Anode of LED is connected to P2.0 through a resistor of 330 Ohm and cathode is connected to GND.
6. For programming user can use hardware programmer to program the 8051 uC or one can also use in system programming.

Figure 14.2 shows the circuit diagram of the system.

14.3   PROGRAM

```
sbit LED=P2^0; //LED at port 2.0

voidmain()
{
       charData_in;
       UART_Init();
       P2 = 0;
       LED = 0;
       while(1)
       {
             Data_in = UART_RxChar();
             if(Data_in == '1')
             {
                    LED = 1; //Turn ON LED
                    UART_SendString("LED_ON");
             }
             elseif(Data_in == '2')
             {
                    LED = 0; // Turn OFF LED
                    UART_SendString("LED_OFF");  // Send status of LED

             }
             else
                    UART_SendString("Select option");

       }
}
```

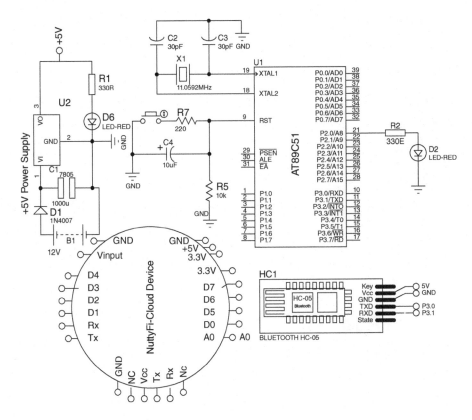

FIGURE 14.2 Circuit diagram of the system.

Section D

Case Study Based on Data
Logger to Cloud Server

15 Health Monitoring System for Solar Panel with Cayenne APP

The solar system inspection is important on regular basis to improve longevity and ensure performance of the solar system. Intelligent monitoring can improve the efficiency of solar panels. This chapter discusses the health monitoring system for solar panel with Cayenne App.

15.1 INTRODUCTION

The system comprises of +12V/500m, a power supply, NuttyFi, Liquid crystal display (LCD) and UV index sensor, voltage sensor, and current sensor. The main objective of the system is to display the solar PV parameters data on LCD by reading the UV index sensor, voltage sensor, and current sensor from solar panel configuration. The sensory information from sensors are sampled and recorded by Arduino. 8051 and NuttyFi/Wi-Fi modem are connected serially. NuttyFi/Wi-Fi modem transfers the sensory data packet to cloud and Cayenne App using Wi-Fi. Figure 15.1 shows the block diagram of the system. Table 15.1 shows the components list required to develop the system.

15.2 CIRCUIT DIAGRAM

Connect the components as follows:

1. Connect +12V power supply output to input of +12V to +5V convertor.
2. Connect +5V output to power up the ADC 0804 board and also connect to LCD to power up.
3. Connect 1, 16 pins of LCD to GND and 2, 15 pins to +5V.
4. Connect the variable terminal of 10K POT to pin 3 of LCD to control the contrast.
5. Control lines RS, RW, and E of LCD are connected to P2.0, P2.1, and P2.2 respectively.
6. Connect D4, D5, D6, and D7 of LCD to P3.4, P3.5, P3.6, and P3.7 pins of 8051 uC respectively.
7. Connect eight output pins of ADC0804 to the P1 port of the 8051 uC.
8. For programming user can use hardware programmer to program the 8051 uC or one can also use in system programming.
9. Connect the control lines C(9), B(10), and C(11) of MUX 74HC4051 to P2.5, P2.6, and P2.7 pins of 8051 uC respectively.
10. Assign pins X0(13), X1(14), X2(15), X3(12), X4(1),X5(5),X6(2), and X7(4) pins of MUX 74HC4051to A0, A1, A2, A3, A4, A5, A6, and A7 respectively. These are useful to connect analog sensor with it.
11. Connect pin 3 of MUX 74HC4051 to Vin+ pins of ADC0804.
12. Connect +Vcc, GND, and OUT pin of voltage sensor to +5V, GND and A0 pin of the MUX 74HC4051. Select the A0 by giving the control signal (ABC=000) to MUX 74HC4051 from 8051 uC.
13. Connect +Vcc, GND, and OUT pin of current sensor to +5V, GND, and A1 pin of the MUX 74HC4051. Select the A1 by giving the control signal (ABC=001) to MUX 74HC4051 from 8051 uC.

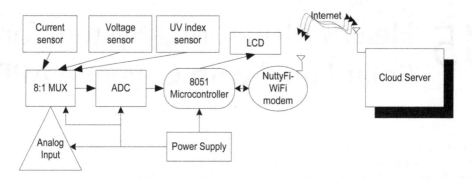

FIGURE 15.1   Block diagram of the system.

14. Connect +Vcc, GND, and OUT pin of UV index sensor to +5V, GND, and A2 pin of the MUX 74HC4051. Select the A2 by giving the control signal (ABC=010) to MUX 74HC4051 from 8051 uC.
15. Connect Vin, GND, Tx and Rx of NuttyFi to +5V, GND, Rx and Tx pins of the 8051 uC respectively.

Figure 15.2 shows the circuit diagram of the system.

15.3   PROGRAM

To connect the system with Wi-Fi network, 8051 and NuttyFi both needs to be programmed separately.

TABLE 15.1
Components List

Component/Specification	Quantity
Power supply 12V/1Amp	1
2 Relay Board	1
Jumper wire M-M	20
Jumper wire M-F	20
Jumper wire F-F	20
Power supply extension (To get more +5V and GND)	1
LCD16*2	1
LCD patch/explorer board	1
NuttyFi patch	1
NuttyFi	1
Voltage sensor	1
Current sensor	1
UV index sensor	1
8051 board with ADC and MUX74HC4051	1

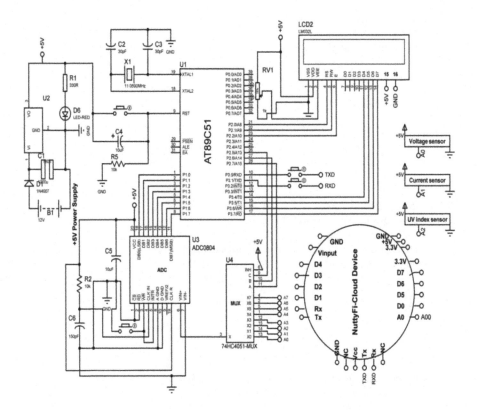

FIGURE 15.2 Circuit diagram of the system.

15.3.1 PROGRAM FOR 8051 MICROCONTROLLER

```
#include<reg52.h>
#include<stdio.h>
#include<string.h>
#include<stdlib.h>
void delay();
void transmit(char);
sbit lcd_E = P2^2;
sbit lcd_RW = P2^1;
sbit lcd_RS = P2^0;

void lcd_init();
void send_command(char command);
void send_data(char databyte);
void delay_lcd();
void display(char, char[]);

char databyte[]={"0123456789"};

char a,b,c;
char x,y,z;
void main(void)
{
```

```
TMOD=0x20; // Timer1 Mode2 8 bit auto reload
TH1=0xfd; // 9600 bps
SCON=0x50; // 8 Data bit, 1 start bit, 1 stop bit
TR1=1; // Timer1 ON
while(1)
        {
                unsigned char i;
                i=P1;
                a=i%10;
                b=i/10;
                c=i/100;
                lcd_init();

                P25=0; P26=0; P27=0;
                display(0x80, "Welcome to IoT ");
                display(0xc0, "Voltage:  ");
                send_command(0xc9);         send_data(databyte[c]);
                send_command(0xca);         send_data(databyte[b]);
                send_command(0xcb);         send_data(databyte[a]);
                delay();
                x=P1;

                P25=0; P26=0; P27=1;
                display(0x80, "Welcome to IoT ");
                display(0xc0, "Current :  ");
                send_command(0xc9);         send_data(databyte[c]);
                send_command(0xca);         send_data(databyte[b]);
                send_command(0xcb);         send_data(databyte[a]);
                delay();
                y=P2;
                P25=0; P26=1; P27=0;
                display(0x80, "Welcome to IoT ");
                display(0xc0, "UV Index:  ");
                send_command(0xc9);         send_data(databyte[c]);
                send_command(0xca);         send_data(databyte[b]);
                send_command(0xcb);         send_data(databyte[a]);
                delay();
                z=P3;

                transmit(x);        transmit(',');        transmit(y);        transmit(',');
                transmit(z);        transmit('\n');
        }
}

void transmit(char y1)
{
        SBUF=y1;
        while(TI==0);
        TI=0;
}

void display(char x, char databyte[])
{
        char y=0;
        send_command(x);
```

```
        while(databyte[y]!='\0')
        {
                send_data(databyte[y]);
                y++;
        }
}

void lcd_init()
{
        send_command(0x03);             delay_lcd();
        send_command(0x03);             delay_lcd();
        send_command(0x03);             delay_lcd();
        send_command(0x03);             delay_lcd();
        send_command(0x03);             delay_lcd();
        send_command(0x03);             delay_lcd();
        send_command(0x28);             delay_lcd();
        send_command(0x28);             delay_lcd();
        send_command(0x0E);             delay_lcd();
        send_command(0x01);             delay_lcd();
        send_command(0x0c);             delay_lcd();
        send_command(0x06);             delay_lcd();
}

void send_command(char command)
{
        char x,y;
        lcd_RS = 0;                     /* (lcd_RS) P2.3---->RS , RS=0 for Instruction Write */
        lcd_RW = 0;                     /* (lcd_RW) P2.2---->R/W, W=0 for write operation */
        lcd_E  = 1;                     /* (lcd_E) P2.5---->E ,E=1 for Instruction execution */

        x=command;
        x=x >> 4;
        P3=P3 & 0xf0;
        P3=P3 | x;

        lcd_E  = 0;                     /* E=0; Disable lcd */
        lcd_RW = 1;                     /* W=1;  Disable write signal */
        lcd_RS = 1;                     /* RS=1 */

        lcd_RS = 0;                     /* (lcd_RS) P2.3---->RS , RS=0 for Instruction Write */
        lcd_RW = 0;                     /* (lcd_RW) P2.2---->R/W, W=0 for write operation */
        lcd_E  = 1;                     /* (lcd_E) P2.5---->E ,E=1 for Instruction execution */

        y=command;
        y=y & 0x0f;
        P2=P2 & 0xf0;
        P2=P2 | y;

        lcd_E  = 0;                     /* E=0; Disable lcd */
        lcd_RW = 1;                     /* W=1;  Disable write signal */
        lcd_RS = 1;                     /* RS=1 */
}

void send_data(char databyte)
{
        char x,y;
```

```
        lcd_RS = 1;                   /* (lcd_RS) P2.3---->RS , RS=1 for Data Write */
        lcd_RW = 0;                   /* (lcd_RW) P2.2---->R/W, W=0 for write operation */
        lcd_E = 1;                    /* (lcd_E) P2.5---->E ,E=1 for Instruction execution */

        x=databyte;
        x=x >> 4;
        P3=P3 & 0xf0;
        P3=P3 | x;

        lcd_E = 0;                    /* E=0; Disable lcd */
        lcd_RW = 1;                   /* W=1;  Disable write signal */
        lcd_RS = 0;                   /* RS=0 */

        lcd_RS = 1;                   /* (lcd_RS) P2.3---->RS , RS=1 for Data Write */
        lcd_RW = 0;                   /* (lcd_RW) P2.2---->R/W, W=0 for write operation */
        lcd_E = 1;                    /* (lcd_E) P2.5---->E ,E=1 for Instruction execution */

        y=databyte;
        y=y & 0x0f;
        P2=P2 & 0xf0;
        P2=P2 | y;

        lcd_E = 0;                    /* E=0; Disable lcd */
        lcd_RW = 1;                   /* W=1;  Disable write signal */
        lcd_RS = 0;                   /* RS=0 */
}

void delay_lcd()
{
        int j;
              for(j=0;j<100;j++)
              {
              }
}

void delay()
{
        int i,j;
        for(i=0;i<=75;i++)
        {
              for(j=0;j<=2500;j++)
              {}
        }
}
```

15.3.2 Program for NuttyFi

```
        #define CAYENNE_PRINT Serial
        #include <CayenneMQTTESP8266.h>
        #include "StringSplitter.h"
        char ssid[] = "ESPServer"; // add hotspot id here
        char wifiPassword[] = "DDDD@12345"; // add hotspot password here

        char username[] = "fac81bb0-7283-11e7-85a3-9540e9f7b5aa";
        char password[] = "3745eb389f4e035711428158f7cdc1adc0475946";
```

```
char clientID[] = "386b86f0-7284-11e7-b0bc-87cd67a1f8c7";

unsigned long lastMillis = 0;
String VOLT,AMP, UV;
String inputString_NUTTY= "";
void setup()
{
        Serial.begin(9600); // initialize serial communication
        Cayenne.begin(username, password, clientID, ssid, wifiPassword);// start cayenne app
}

void loop() {
        Cayenne.loop();
  serialEvent_NODEMCU() ;
  if (millis() - lastMillis > 10000)
{
  lastMillis = millis();
Cayenne.virtualWrite(0, UV);
   Cayenne.virtualWrite(2, VOLT);
   Cayenne.virtualWrite(1,AMP);
}
}

CAYENNE_IN_DEFAULT()
{
        CAYENNE_LOG("CAYENNE_IN_DEFAULT(%u) - %s, %s", request.channel, getValue.
getId(), getValue.asString());
}

void serialEvent_NUTTY()
{

  while (Serial.available()>0)
  {
  inputString_NODEMCU = Serial.readStringUntil('\n');// Get serial input
  StringSplitter *splitter = new StringSplitter(inputString_NUTTY, ',', 2); // new
StringSplitter(string_to_split, delimiter, limit)
  int itemCount = splitter->getItemCount();

  for(int i = 0; i < itemCount; i++)
   {
   String item = splitter->getItemAtIndex(i);
   VOLT = splitter->getItemAtIndex(0);
   AMP = splitter->getItemAtIndex(1);
   UV = splitter->getItemAtIndex(2);

    delay(200);
   }
   inputString_NUTTY = "";
  }

}
```

```
//#define CAYENNE_DEBUG
#define CAYENNE_PRINT Serial
#include <CayenneMQTTESP8266.h>
#include <ESP8266WiFi.h>
#include <SoftwareSerial.h>
SoftwareSerial mySerial(D7,D8,false,256);
// WiFi network info.
char ssid[] = "ESPServer_RAJ";
char wifiPassword[] = "RAJ@12345";

// Cayenne authentication info. This should be obtained from the Cayenne Dashboard.
char username[] = "fac81bb0-7283-11e7-85a3-9540e9f7b5aa";
char password[] = "3745eb389f4e035711428158f7cdc1adc0475946";
char clientID[] = "386b86f0-7284-11e7-b0bc-87cd67a1f8c7";
```

FIGURE 15.3   Snapshot showing username, password to MQTT.

15.4   CLOUD SERVER

Steps to add NuttyFi in Cayenne cloud:

1. Install the Arduino IDE and add Cayenne MQTT Library to Arduino IDE.
2. Install the ESP8266 board package to Arduino IDE.
3. Install required USB driver on computer to program the ESP8266.
4. Connect the ESP8266 to PC/Mac via data-capable USB cable.
5. In the Arduino IDE, go to the **tools** menu, select the **board**, and now the **port** ESP8266 is connected to.
6. Use the MQTT username, MQTT password, client ID as well as ssid[], and wifi-Passord[] in the Arduino IDE to write code, as shown in Figure 15.3.
7. Burn the code in 8051 and NuttyFi then window will open; Figure 15.4 shows the snapshots for the developed mobile app after burning program.
8. Figure 15.5 shows the front end of App, showing the data on channel.

FIGURE 15.4   Cayenne App.

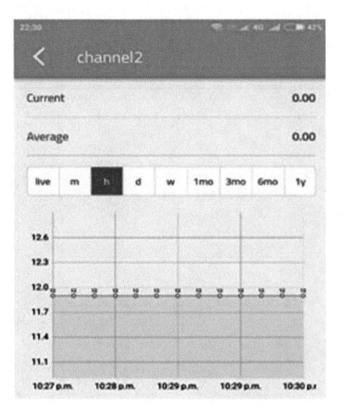

FIGURE 15.5 Sensory data on App.

16 Smart Irrigation System in Agricultural Field with Blynk APP

Smart irrigation system is an Internet of things (IoT) based device, capable of automatic irrigation and field monitoring system. Automation can be achieved with the help of monitoring the field and control with respect to the sensory data received. This chapter discusses the smart irrigation system in agricultural field with Blynk App.

16.1 INTRODUCTION

The objective of the system is to develop a smart control for site-specific management of irrigation system with Blynk App. Blynk mobile app is designed to control the water management in the field as per requirement. The complete system comprises of two sections—field device and mobile app. Field device comprises of 8051, NuttyFi, power supply, LCD, relay board, ADC 0804, MUX 74HC4051, soil moisture sensor, temperature and humidity sensor, water level sensor, motor IN, and motor OUT. The system is designed to establish control for specific agricultural field by taking sensory data from the sensors and control the PUMP IN motor and PUMP OUT motor. It is done with the help of mobile app. Figure 16.1 shows the block diagram of the system. Table 16.1 shows the list of components required to develop the system.

16.2 CIRCUIT DIAGRAM

The connections of circuit are as follows:

1. Connect **SOIL sensor** output pin OUTPUT_SS to pinA0 of MUX 74HC4051.
2. Connect +Vcc and GND pins of SOIL sensor to +5V and GND.
3. Connect **Ultrasonic sensor** pin trigger and ECHO to pin P3.2 and P3.3 of 8051.
4. Connect +Vcc and GND pins of ultrasonic sensor to +5V and GND of power supply
5. Connect **DHT 11** pin 2 to pin P2.3 of 8051.
6. Connect +5V output to power up the ADC 0804 board and also connect to LCD to power up.
7. Connect 1, 16 pins of LCD to GND and 2, 15 pins to +5V.
8. Connect the variable terminal of 10K POT to pin 3 of LCD to control the contrast.
9. A control line like RS, RW, and E of LCD is connected to P2.0, P2.1, and P2.2 respectively.
10. Connect D4, D5, D6, and D7 of LCD to P3.4, P3.5, P3.6, and P3.7 pins of 8051 uC respectively.
11. Connect +Vcc, GND, and OUT pin of the gas sensor to +5V, GND, and input pins of the ADC 0804.
12. Connect 8 output pin of ADC0804 to the P1 port of the 8051 uC.
13. For programming user can use hardware programmer to program the 8051 uC or one can also use in system programming.
14. Connect the control lines C(9), B(10), and C(11) of MUX 74HC4051 to P2.5, P2.6, and P2.7 pins of 8051 uC respectively.
15. Assign pin X0(13), X1(14), X2(15), X3(12), X4(1),X5(5),X6(2), and X7(4) of MUX 74HC4051to A0, A1, A2, A3, A4, A5, A6, and A7 respectively. These are useful to connect analog sensors with it.

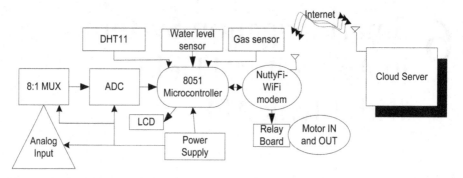

FIGURE 16.1   Block diagram of the system.

16. Connect pin 3 of MUX 74HC4051 to Vin+ pins of ADC0804.
17. Connect V_{IN}, GND, Tx, and Rx of NuttyFi to +5V, GND, Rx, and Tx pins of the 8051 uC.
18. Water Pump IN motor and Water Pump OUT motor to D3and D4 pins of NuttyFi using relay board.
19. The base of NPN transistor 2N2222 is to be connected with pin D3 and D4 of NuttyFi. Emitter of transistor is grounded.
20. Collector of transistor is to be connected with L2 of relay and Li of relay to positive terminal of 12V battery.
21. Negative terminal of battery is connected with ground.
22. One terminal of appliance (pump motor) is connected with "NO" of relay and other to one end the AC source.
23. Other end of AC source is connected to "Common" terminal of relay.
24. Connect +Vcc, GND, and OUT pin of MQ2 gas sensor to +5V, GND, and A0 pin of the MUX 74HC4051. Select the A0 by giving the control signal (ABC=000) to MUX 74HC4051 from 8051 uC.
25. Connect +Vcc, GND, Trig, and Echo pin of HC05 sensor to +5V, GND, P2.3, and P2.4 pin of the 8051 respectively.
26. Connect +Vcc, GND, 2 pin of DHT11 sensor to +5V, GND, and P3.3 pin of the 8051 respectively.

Figure 16.2 shows the circuit diagram of the system.

TABLE 16.1
Components List

Component/Specification	Quantity
Power supply 12V/1Amp	1
2 Relay Board	1
Jumper wire M-M	20
Jumper wire M-F	20
Jumper wire F-F	20
Power supply extension (To get more +5V and GND)	1
LCD20*4	1
LCD patch/explorer board	1
NuttyFi patch	1
NuttyFi	1
Soil moisture sensor	1
Ultrasonic sensor patch	1
DHT11	1
8051 board with ADC and MUX74HC4051	1

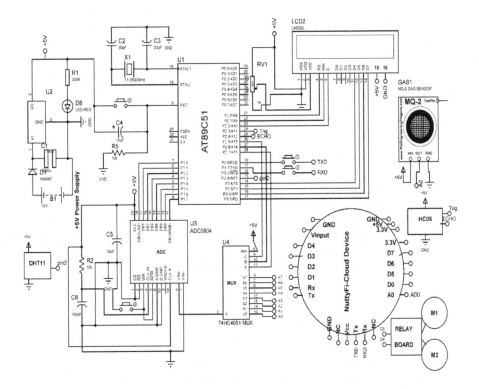

FIGURE 16.2 Circuit diagram of the system.

16.3 PROGRAM

16.3.1 PROGRAM FOR 8051

```c
#include<reg52.h>
#include<stdio.h>
#include<string.h>
#include<stdlib.h>

void delay();
void transmit(char);
sbit lcd_E = P2^2;
sbit lcd_RW = P2^1;
sbit lcd_RS = P2^0;

void lcd_init();
void send_command(char command);
void send_data(char databyte);
void delay_lcd();
void display(char, char[]);

char databyte[]={"0123456789"};

char a,b,c;
char x,y,z;
void main(void)
{
TMOD=0x20; // Timer1 Mode2 8 bit auto reload
```

```
TH1=0xfd; // 9600 bps
SCON=0x50; // 8 Data bit, 1 start bit, 1 stop bit
TR1=1; // Timer1 ON
while(1)
        {
                unsigned char i;
                i=P1;
                a=i%10;
                b=i/10;
                c=i/100;
                lcd_init();

                P25=0; P26=0; P27=0;
                display(0x80, "Welcome to IoT ");
                display(0xc0, "Gas Sensor :  ");
                send_command(0xc9);          send_data(databyte[c]);
                send_command(0xca);          send_data(databyte[b]);
                send_command(0xcb);          send_data(databyte[a]);
                delay();
                x=P1;

                P25=0; P26=0; P27=1;
                display(0x80, "Welcome to IoT ");
                display(0xc0, "UltraSonic:  ");
                send_command(0xc9);          send_data(databyte[c]);
                send_command(0xca);          send_data(databyte[b]);
                send_command(0xcb);          send_data(databyte[a]);
                delay();
                y=P2;
                transmit(x);        transmit(',');        transmit(y);        transmit('\n');
        }
}

void transmit(char y1)
{
        SBUF=y1;
        while(TI==0);
        TI=0;
}

void display(char x, char databyte[])
{
        char y=0;
        send_command(x);
        while(databyte[y]!='\0')
        {
                send_data(databyte[y]);
                y++;
        }
}
void lcd_init()
{
        send_command(0x03);                  delay_lcd();
        send_command(0x03);                  delay_lcd();
        send_command(0x03);                  delay_lcd();
        send_command(0x03);                  delay_lcd();
```

```
        send_command(0x03);              delay_lcd();
        send_command(0x03);              delay_lcd();
        send_command(0x28);              delay_lcd();
        send_command(0x28);              delay_lcd();
        send_command(0x0E);              delay_lcd();
        send_command(0x01);              delay_lcd();
        send_command(0x0c);              delay_lcd();
        send_command(0x06);              delay_lcd();
}

void send_command(char command)
{
        char x,y;
        lcd_RS = 0;              /* (lcd_RS) P2.3---->RS , RS=0 for Instruction Write */
        lcd_RW = 0;              /* (lcd_RW) P2.2---->R/W, W=0 for write operation */
        lcd_E  = 1;              /* (lcd_E)  P2.5---->E ,E=1 for Instruction execution */

        x=command;
        x=x >> 4;
        P3=P3 & 0xf0;
        P3=P3 | x;
        lcd_E  = 0;              /* E=0; Disable lcd */
        lcd_RW = 1;              /* W=1;  Disable write signal */
        lcd_RS = 1;              /* RS=1 */
        lcd_RS = 0;              /* (lcd_RS) P2.3---->RS , RS=0 for Instruction Write */
        lcd_RW = 0;              /* (lcd_RW) P2.2---->R/W, W=0 for write operation */
        lcd_E  = 1;              /* (lcd_E)  P2.5---->E ,E=1 for Instruction execution */
        y=command;
        y=y & 0x0f;
        P2=P2 & 0xf0;
        P2=P2 | y;
        lcd_E  = 0;              /* E=0; Disable lcd */
        lcd_RW = 1;              /* W=1;  Disable write signal */
        lcd_RS = 1;              /* RS=1 */
}

void send_data(char databyte)
{
        char x,y;
        lcd_RS = 1;              /* (lcd_RS) P2.3---->RS , RS=1 for Data Write */
        lcd_RW = 0;              /* (lcd_RW) P2.2---->R/W, W=0 for write operation */
        lcd_E  = 1;              /* (lcd_E) P2.5---->E ,E=1 for Instruction execution */
        x=databyte;
        x=x >> 4;
        P3=P3 & 0xf0;
        P3=P3 | x;

        lcd_E = 0;               /* E=0; Disable lcd */
        lcd_RW = 1;              /* W=1;  Disable write signal */
        lcd_RS  = 0;             /* RS=0 */

        lcd_RS = 1;              /* (lcd_RS) P2.3---->RS , RS=1 for Data Write */
        lcd_RW = 0;              /* (lcd_RW) P2.2---->R/W, W=0 for write operation */
        lcd_E  = 1;              /* (lcd_E)  P2.5---->E ,E=1 for Instruction execution */

        y=databyte;
```

```
        y=y & 0x0f;
        P2=P2 & 0xf0;
        P2=P2 | y;

        lcd_E = 0;              /* E=0; Disable lcd */
        lcd_RW = 1;             /* W=1;  Disable write signal */
        lcd_RS = 0;             /* RS=0 */
}

void delay_lcd()
{
        int j;
                for(j=0;j<100;j++)
                {
                }
}

void delay()
{
        int i,j;
        for(i=0;i<=75;i++)
        {
                for(j=0;j<=2500;j++)
                {}
        }
}
```

16.3.2 Program for NuttyFi

```
        #include "StringSplitter.h"
        #define BLYNK_PRINT Serial
        //////// library for NodeMCU
        #include <ESP8266WiFi.h>
        #include <BlynkSimpleEsp8266.h>
        char auth[] = "5c8e33bf09a04b03b2fa153928b035c5";///add token here
        char ssid[] = "ESPServer "; // SSID of hotspot
        char pass[] = "CCCC@12345"; // Password of hotspot
        ///////// library for internal LCD of blynk APP
        WidgetLCD LCD_BLYNK(V8);
        ///// for timer
        BlynkTimer timer;

        int PUMP_IN=D3;//Connect relay input to D3 of motor IN
        int PUMP_OUT=D4;// Connect relay input to D4 of motor OUT
        String ULTRA,TEMP,HUM,SOIL;
         String CONT_NEW_STRING= "";
        ///////////////////// use button
        BLYNK_WRITE(V1)
        {
         int PUMP_IN_VAL = param.asInt();
         if(PUMP_IN_VAL==HIGH)
         {
          lcd.clear(); // clear the contents of LCD
          digitalWrite(PUMP_IN,HIGH); // Make D3 to HIGH
          digitalWrite(PUMP_OUT,LOW); // Make D4 to LOW
          ////// external LCD with NOdeMCU
```

```
      lcd.setCursor(0,0); // set cursor on LCD
      lcd.print("PUMP_In Tigger");
      //// LCD blynk
      LCD_BLYNK.print(0,0,"PUMP_In Tigger"); // print string on Blynk LCD
      delay(10); // delay of 10 mSec
      }

}

BLYNK_WRITE(V2)
{
  int PUMP_OUT_VAL = param.asInt();
  if(PUMP_OUT_VAL==HIGH)
  {
    lcd.clear();
    digitalWrite(PUMP_IN,LOW); // make D3 pin to LOW
    digitalWrite(PUMP_OUT,HIGH); // Make D4 pin to HIGH
    LCD_BLYNK.print(0,0,"PUMP_OUT Tigger"); // print string on blynk LCD
    delay(10); // wait for 10 mSec
  }
}
BLYNK_WRITE(V3)
{
  int BOTH_ON = param.asInt();
  if(BOTH_ON==HIGH)
  {
    lcd.clear(); // Clear the contents of LCD
    digitalWrite(PUMP_IN,HIGH); // make D3 pin to HIGH
    digitalWrite(PUMP_OUT,HIGH); // make D4 pin to HIGH
    LCD_BLYNK.print(0,0,"BOTH ON"); // print string on Blynk APP
    delay(10); // wait for 10 mSec
  }
}
/////// Function to read analog sensor
void READ_SENSOR()
{
  serialEvent_NODEMCU(); // call serial event to read serial data from 8051
  Blynk.virtualWrite(V4,SOIL); // Print value on virtual pin
  Blynk.virtualWrite(V5,ULTRA); // Print value on virtual pin

  Blynk.virtualWrite(V6,TEMP); // Print value on virtual pin
  Blynk.virtualWrite(V7,HUM); //Print value on virtual pin
}

void setup()
{
  Serial.begin(9600); // initialize serial communication
  lcd.begin(20, 4); // initialize LCD
  Blynk.begin(auth, ssid, pass);
  pinMode(PUMP_IN,OUTPUT);//D3 pin of Nuttyfi
  pinMode(PUMP_OUT,OUTPUT);//D4 pin of Nuttyfi
  timer.setInterval(1000L,READ_SENSOR);//// read sensor with setting delay of 1 Sec
}

void loop()
{
```

```
    Blynk.run(); // initial blynk run
    timer.run(); // Initiates BlynkTimer
}


void serialEvent_NODEMCU()
{
  while (Serial.available()>0) // check serial data availability
  {
  CONT_NEW_STRING = Serial.readStringUntil('\n');// Get serial input
StringSplitter *splitter = new StringSplitter(CONT_NEW_STRING, ',', 4);  // new
StringSplitter(string_to_split, delimiter, limit)
int itemCount = splitter->getItemCount();
for(int i = 0; i < itemCount; i++)
   {
     String item = splitter->getItemAtIndex(i);
     SOIL = splitter->getItemAtIndex(0); // get the first value from 8051
     ULTRA = splitter->getItemAtIndex(1); // get the second value from 8051
     TEMP = splitter->getItemAtIndex(2); // get the third value from 8051
     HUM= splitter->getItemAtIndex(3); // get the fourth value from 8051
   }
  CONT_NEW_STRING= "";
  delay(20); // wait for 20 mSec
  }
}
```

16.4   CLOUD SERVER

Blynk is iOS and Android platform to design mobile app. To design the app download latest Blynk library from:https://github.com/blynkkk/blynk-library/releases/latest

Mobile App can easily be designed just bydragging and dropping widgets on the provided space. Tutorials can be downloaded from: http://www.blynk.cc

16.4.1   Steps to Design Blynk App

1. **Step1:** Download and install the Blynk App for your mobile Android or iphone from http://www.blynk.cc/getting-started/
2. **Step2:** Create a Blynk Account
3. **Step3:** Create a new project
     Click on + for creating new project and choose the theme dark (black background) or light (white background) and click on create, as shown in Figure 16.3.
4. **Step4:** Auth token is a unique identifier, which will be received on the email address user, provide at time of making account. Save this token, as this is required to copy in the main program of receiver section.
5. **Step5:** Select the device to which smart phone needs to communicate, e.g., ESP8266 (NodeMCU).
6. **Step6:** Open widget box and select the components required for the project. For this project five buttons are selected.
7. **Step7:**Tap on the widget to get its settings, select virtual terminals as V1, V2 for each buttons, which need to be defined later on the program.
8. **Step8:** After completing the widget settings, Run the project.
9. Front end of App for the system, as shown in Figure 16.4.

FIGURE 16.3 Create a new project.

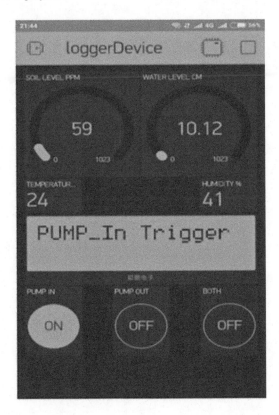

FIGURE 16.4 BlynkApp.

17 Environmental Parameters Monitoring System with Virtuino APP

Environmental parameters monitoring system is capable of sensing different parameters to check the health of environment. This chapter discusses the smart irrigation system in agricultural field with Blynk App.

17.1 INTRODUCTION

Figure 17.1 shows the block diagram of the system. The system comprises of 8051, DC 12V/1Amp adaptor, 12V to 5V, 3.3V converter, BMP180 sensor, MQ135sensor, Dust sampler, liquid crystal display, and NuttyFi. The objective of the system is to display the information of BMP180 sensor, air quality using MQ135, and dust sampler of air on liquid crystal. The sensors are interfaced to 8051. The data packet is formed with 8051, which contains all sensory information. The NuttyFi/Wi-Fi modem transfers the packet information to cloud APP/cloud server. Table 17.1 shows the list of components required to develop the system.

17.2 CIRCUIT DIAGRAM

Circuit connections are as follows:

1. Connect +12V power supply output to input of +12V to +5V convertor.
2. Connect +5V output to power up the ADC 0804 board and also connect to LCD to power up.
3. Connect 1, 16 pins of LCD to GND and 2, 15 pins to +5V.
4. Connect the variable terminal of 10K POT to pin 3 of LCD to control the contrast.
5. A control line like RS, RW, and Enable of LCD is connected to P2.0, P2.1, and P2.2 respectively.
6. Connect D4, D5, D6, and D7 of LCD to P3.4, P3.5, P3.6, and P3.7 pins of 8051 uC respectively.
7. Connect +Vcc, GND, and OUT pin of the gas sensor to +5V, GND and input pins of the ADC 0804.
8. Connect 8 output pin of ADC0804 to the P1 port of the 8051 uC.
9. For programming user can use hardware programmer to program the 8051 uC or one can also use in system programming.
10. Connect the control lines C(9), B(10), and C(11) of MUX 74HC4051 to P2.5, P2.6, and P2.7 pins of 8051 uC respectively.
11. Assign pin X0(13), X1(14), X2(15), X3(12), X4(1),X5(5),X6(2), and X7(4) pins of MUX 74HC4051to A0, A1, A2, A3, A4, A5, A6, and A7 respectively. These are useful to connect analog sensor with it.
12. Connect pin 3 of MUX 74HC4051 to Vin+ pins of ADC0804.

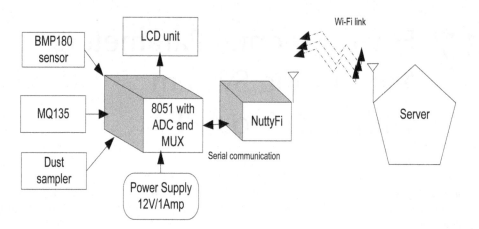

FIGURE 17.1   Circuit diagram of the system.

13. Connect +Vcc, GND, and OUT pin of MQ135 sensor to +5V, GND, and A0 pin of the MUX 74HC4051. Select the A0 by giving the control signal (ABC=000) to MUX 74HC4051 from 8051 uC.
14. Connect +Vcc, GND, and OUT pin of Dust sampler sensor to +5V, GND, and A1 pin of the MUX 74HC4051. Select the A1 by giving the control signal (ABC=001) to MUX 74HC4051 from 8051 uC.
15. Connect +Vcc, GND, and OUT pin of BMP sensor to +5V, GND, and A2 pin of the MUX 74HC4051. Select the A2 by giving the control signal (ABC=010) to MUX 74HC4051 from 8051 uC.
16. Connect V_{IN}, GND, Tx, and Rx of NuttyFi to +5V, GND, Rx, and Tx pins of the 8051 uC.

Figure 17.2 shows the circuit diagram of the system.

TABLE 17.1

Components List

Components	Quantity
LCD20*4	1
LCD20*4 patch	1
DC 12V/1Amp adaptor	1
12V to 5V, 3.3V converter	1
LED with 330 Ohm resistor	1
Jumper wire M to M	20
Jumper wire M to F	20
Jumper wire F to F	20
BMP180 sensor	1
MQ135sensor	1
Dust sampler	1
8051 board with ADC0804 and MAX74HC4051	1
NuttyFi	1
NuttyFi breakout board/Patch	1

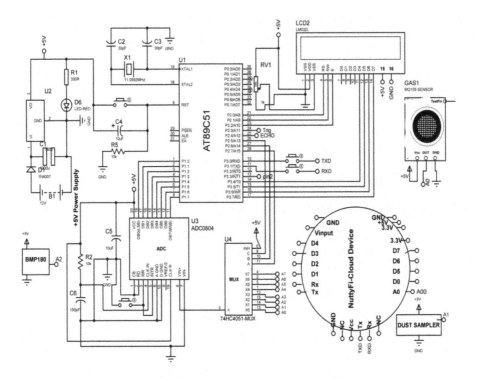

FIGURE 17.2   Circuit diagram of the system.

17.3   PROGRAM

17.3.1   Program for 8051

```
#include<reg52.h>
#include<stdio.h>
#include<string.h>
#include<stdlib.h>

void delay();
void transmit(char);
sbit lcd_E  = P2^2;
sbit lcd_RW = P2^1;
sbit lcd_RS = P2^0;

void lcd_init();
void send_command(char command);
void send_data(char databyte);
void delay_lcd();
void display(char, char[]);

char databyte[]={"0123456789"};

char a,b,c;
char x,y,z;
void main(void)
{
TMOD=0x20; // Timer1 Mode2 8 bit auto reload
TH1=0xfd; // 9600 bps
```

```
SCON=0x50; // 8 Data bit, 1 start bit, 1 stop bit
TR1=1; // Timer1 ON
while(1)
        {
                unsigned char i;
                i=P1;
                a=i%10;
                b=i/10;
                c=i/100;
                lcd_init();

                P25=0; P26=0; P27=0;
                display(0x80, "Welcome to IoT ");
                display(0xc0, "Gas Sensor :  ");
                send_command(0xc9);          send_data(databyte[c]);
                send_command(0xca);          send_data(databyte[b]);
                send_command(0xcb);          send_data(databyte[a]);
                delay();
                x=P1;
                transmit(x);          transmit('\n');
        }
}

void transmit(char y1)
{
        SBUF=y1;
        while(TI==0);
        TI=0;
}

void display(char x, char databyte[])
{
        char y=0;
        send_command(x);
        while(databyte[y]!='\0')
        {
                send_data(databyte[y]);
                y++;
        }
}

void lcd_init()
{
        send_command(0x03);          delay_lcd();
        send_command(0x03);          delay_lcd();
        send_command(0x03);          delay_lcd();
        send_command(0x03);          delay_lcd();
        send_command(0x03);          delay_lcd();
        send_command(0x03);          delay_lcd();
        send_command(0x28);          delay_lcd();
        send_command(0x28);          delay_lcd();
        send_command(0x0E);          delay_lcd();
        send_command(0x01);          delay_lcd();
        send_command(0x0c);          delay_lcd();
        send_command(0x06);          delay_lcd();
}
```

```
void send_command(char command)
{
        char x,y;
        lcd_RS = 0;              /* (lcd_RS) P2.3---->RS , RS=0 for Instruction Write */
        lcd_RW = 0;              /* (lcd_RW) P2.2---->R/W, W=0 for write operation */
        lcd_E  = 1;              /* (lcd_E)  P2.5---->E ,E=1 for Instruction execution */

        x=command;
        x=x >> 4;
        P3=P3 & 0xf0;
        P3=P3 | x;

        lcd_E  = 0;              /* E=0; Disable lcd */
        lcd_RW = 1;              /* W=1;  Disable write signal */
        lcd_RS = 1;              /* RS=1 */

        lcd_RS = 0;              /* (lcd_RS) P2.3---->RS , RS=0 for Instruction Write */
        lcd_RW = 0;              /* (lcd_RW) P2.2---->R/W, W=0 for write operation */
        lcd_E  = 1;              /* (lcd_E)  P2.5---->E ,E=1 for Instruction execution */

        y=command;
        y=y & 0x0f;
        P2=P2 & 0xf0;
        P2=P2 | y;

        lcd_E  = 0;              /* E=0; Disable lcd */
        lcd_RW = 1;              /* W=1;  Disable write signal */
        lcd_RS = 1;              /* RS=1 */
}

void send_data(char databyte)
{
        char x,y;
        lcd_RS = 1;              /* (lcd_RS) P2.3---->RS , RS=1 for Data Write */
        lcd_RW = 0;              /* (lcd_RW) P2.2---->R/W, W=0 for write operation */
        lcd_E = 1;               /* (lcd_E) P2.5---->E ,E=1 for Instruction execution */

        x=databyte;
        x=x >> 4;
        P3=P3 & 0xf0;
        P3=P3 | x;

        lcd_E = 0;               /* E=0; Disable lcd */
        lcd_RW = 1;              /* W=1;  Disable write signal */
        lcd_RS  = 0;             /* RS=0 */

        lcd_RS = 1;              /* (lcd_RS) P2.3---->RS , RS=1 for Data Write */
        lcd_RW = 0;              /* (lcd_RW) P2.2---->R/W, W=0 for write operation */
        lcd_E  = 1;              /* (lcd_E) P2.5---->E ,E=1 for Instruction execution */

        y=databyte;
        y=y & 0x0f;
        P2=P2 & 0xf0;
        P2=P2 | y;

        lcd_E = 0;               /* E=0; Disable lcd */
```

```
        lcd_RW = 1;              /* W=1;  Disable write signal */
        lcd_RS  = 0;             /* RS=0 */
}

void delay_lcd()
{
        int j;
                for(j=0;j<100;j++)
                {
                }
}

void delay()
{
        int i,j;
        for(i=0;i<=75;i++)
        {
                for(j=0;j<=2500;j++)
                {}
        }
}
```

17.3.2 PROGRAM FOR NUTTYFI

```
#include <ESP8266WiFi.h>
#include "Virtuino_ESP_WifiServer.h"
const char* ssid = "ESPServer";
const char* password = "FFFF@12345";
WiFiServer server(8000);                   // Server port
Virtuino_ESP_WifiServer virtuino(&server);
int storedValue=0;
int counter =0;
long storedTime=0;
String PRESS,ALT,DUST,MQ135;
String inputString_NODEMCU = "";
void setup()
{
 virtuino.DEBUG=true;                      // set this value TRUE to enable the serial monitor status
 virtuino.password="1234";                 // Set a password to your web server for more protection
 Serial.begin(9600);                       // Enable this line only if DEBUG=true
 delay(10);
 pinMode(2, OUTPUT);
 digitalWrite(2, 0);

 Serial.println("Connecting to "+String(ssid));
 WiFi.mode(WIFI_STA);                      // Config module as station only.
 WiFi.begin(ssid, password);
 while (WiFi.status() != WL_CONNECTED)
  {
   delay(500);
   Serial.print(".");
  }
  Serial.println("");
  Serial.println("WiFi connected");
  Serial.println(WiFi.localIP());
```

```
  server.begin();
  Serial.println("Server started");

}

void loop()
{
  virtuino.run();
  serialEvent_NODEMCU();
   int v1=virtuino.vDigitalMemoryRead(0);          // Read virtual memory 0 from Virtuino app
   int v2=virtuino.vDigitalMemoryRead(1);          // Read virtual memory 0 from Virtuino app
   if (v1!=storedValue)
   {
   Serial.println("-------Virtual pin DV0 is changed to="+String(v1));
   if (v1==1) digitalWrite(D4,0);
   else digitalWrite(D4,1);
   storedValue=v1;
   }
   if (v2!=storedValue)
   {
   Serial.println("-------Virtual pin DV0 is changed to="+String(v2));
   if (v2==1) digitalWrite(D4,0);
   else digitalWrite(D4,1);
   storedValue=v2;
   }
    serialEvent_NODEMCU();
    virtuino.vMemoryWrite(2,PRESS);
    virtuino.vMemoryWrite(3,ALT);
    virtuino.vMemoryWrite(4,DUST);
    virtuino.vMemoryWrite(5,MQ135);
    long t= millis();
   if (t>storedTime+5000)
   {
    counter++;
    if (counter>20) counter=0;    // limit = 20
    storedTime = t;
    virtuino.vMemoryWrite(12,counter);     // write counter to virtual pin V12
   }
  }

void serialEvent_NODEMCU()
{

  while (Serial.available()>0)
  {
  inputString_NODEMCU = Serial.readStringUntil('\n');// Get serial input
  StringSplitter *splitter = new StringSplitter(inputString_NODEMCU, ',',5); // new
  StringSplitter(string_to_split, delimiter, limit)
  int itemCount = splitter->getItemCount();
  for(int i = 0; i < itemCount; i++)
   {
    String item = splitter->getItemAtIndex(i);
    PRESS= splitter->getItemAtIndex(0);
    ALT= splitter->getItemAtIndex(1);
    DUST= splitter->getItemAtIndex(2);
```

```
        MQ135= splitter->getItemAtIndex(3);

      }
      inputString_NODEMCU = "";
      delay(200);
      }
    }
```

17.4   CLOUD SERVER

Virtuino application is human machine interface platform on cloud. Virtuino application is controlled through Bluetooth, Wi-Fi, GPRS, and ThingSpeak.

Steps to develop App are as follows:

1. **Step 1:** Make connection among 8051, NodeMCU, and external devices.
2. **Step 2:** Follow the following steps:
   a. Click to download the Virtuino Library ver 1.1
   b. Run to add the library via Arduino IDE
   c. Burn the hex code in NodeMCU
1. **Step 3:** Add Wi-Fi settings with android device.
2. **Step 4:** Make the application in Virtuino App and run it to interact with ESP8266/NodeMCU.

Figure 17.3 shows the Virtuino App for the system.

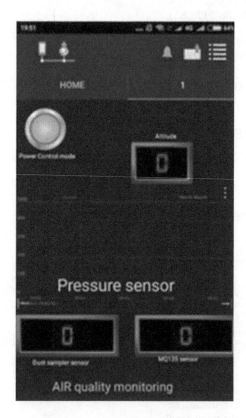

FIGURE 17.3   Virtuino App.

18 Weather Station with ThingSpeak Server

The weather station measures temperature, wind speed, precipitation, atmospheric pressure, and other environmental parameters. Intelligent weather station means it collects data without human intervention. This chapter discusses development of weather station with ThingSpeak server.

18.1 INTRODUCTION

The complete process can be understood with the help of system for the weather station. Internet of things (IoT) makes it intelligent. The atmospheric pressure sensor, temperature sensor, wind speed sensor, and precipitation sensor are used to make intelligent weather station using 8051 and NuttyFi. Along with these sensors, the system comprises of 12V/1Amp DC adaptor, 12V to 5V, 3.3V converter, and liquid crystal display (LCD). The objective of the system is to measure the atmospheric parameters and display the information on LCD. The data packet is formed with Arduino, received from different sensors. Then this data packet is communicated serially to NuttyFi. The NuttyFi, further extract the sensory information and transfer it to cloud. Figure 18.1 shows the block diagram of the system.

Table 18.1 shows the list of components to design a system.

18.1.1 WIND SPEED SENSOR

The **Davis Anemometer** has wind vane with 18k linear potentiometer attached to it. The output from the wind direction circuit can be connected to analog pin on the controller. The controller with a 10-bit analog to digital converter (ADC) gives digital levels variation from 0 to 1023 corresponding to a voltage of 0 to 5V. The mapping for voltage to degree of direction (0–1023 to a 0–360 range) can be done with programming.

18.1.2 CALIBRATION

The simplest way to set up the anemometer for wind direction calibration is to have the mounting arm pointing directly to north on the compass, which means the direction value will line up correctly with North. However, if it is not possible to point the mounting arm to magnetic north, then apply an offset to our wind direction calculation to make correct value of the wind direction reading. If the magnetic north heading relative is between 0 and 180, then subtract the offset from the Direction output and if the magnetic north heading relative is between 181 and 360, then add the offset to the Direction output to get the adjusted wind direction.

18.2 CIRCUIT DIAGRAM

To design the system, connect the components as follows:

1. Connect +12V power supply output to input of +12V to +5V convertor.
2. Connect +5V output to power up the ADC 0804 board and also connect to LCD to power up.
3. Connect 1, 16 pins of LCD to GND and 2, 15 pins to +5V.
4. Connect the variable terminal of 10K POT to pin 3 of LCD to control the contrast.

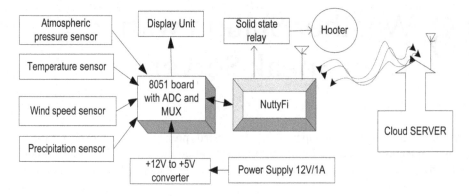

FIGURE 18.1   Block diagram of the system.

5. A control line like RS, RW, and E of LCD is connected to P2.0, P2.1, and P2.2 respectively.
6. Connect D4, D5, D6, and D7 of LCD to P3.4, P3.5, P3.6, and P3.7 pins of 8051 uC respectively.
7. Connect +Vcc, GND, and OUT pin of the gas sensor to +5V, GND, and input pins of the ADC 0804.
8. Connect eight output pin of ADC0804 to the P1 port of the 8051 uC.
9. For programming user can use hardware programmer to program the 8051 uC or one can also use in system programming.
10. Connect the control lines C(9), B(10), and C(11) of MUX 74HC4051 to P2.5, P2.6, and P2.7 pins of 8051 uC respectively.
11. Assign pin X0(13), X1(14), X2(15), X3(12), X4(1), X5(5), X6(2), and X7(4) pins of MUX 74HC4051 to A0, A1, A2, A3, A4, A5, A6, and A7 respectively. These are useful to connect analog sensor with it.
12. Connect pin 3 of MUX 74HC4051 to Vin+ pins of ADC0804.

TABLE 18.1
Components List

Sr. No.	Name of Components	Quantity
1	8051 board with ADC and MUX74HC4051	1
2	LCD18*4	1
3	LCD18*4 patch	1
4	DC 12V/1Amp adaptor	1
5	12V to 5V, 3.3V converter	1
6	LED with 330 Ohm resistor	1
7	Jumper wire M to M	18
8	Jumper wire M to F	18
9	Jumper wire F to F	18
10	Atmospheric pressure sensor	1
11	Temperature sensor	1
12	Wind speed sensor	1
13	Precipitation sensor	1
14	NuttyFi	1
15	Breakout board for NuttyFi	1

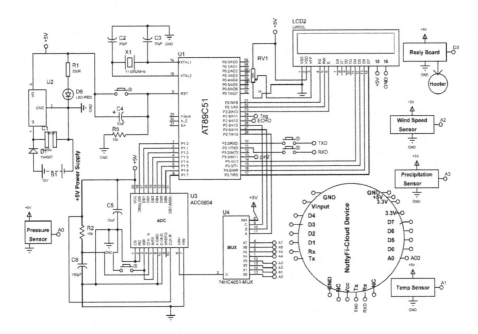

FIGURE 18.2 Circuit diagram of the system.

13. Connect V_{IN}, GND, Tx, and Rx of NuttyFi to +5V, GND, Rx, and Tx pins of the 8051 uC respectively.
14. Connect +Vcc, GND, and OUT pin of **Atmospheric pressure sensor** to +5V, GND, and A0 pin of the MUX 74HC4051. Select the A0 by giving the control signal (ABC=000) to MUX 74HC4051 from 8051 uC.
15. Connect +Vcc, GND and OUT pin of **Temperature sensor** to +5V, GND and A1 pin of the MUX 74HC4051. Select the A1 by giving the control signal (ABC=001) to MUX 74HC4051 from 8051 uC.
16. Connect +Vcc, GND, and OUT pin of **wind speed sensor** to +5V, GND, and A2 pin of the MUX 74HC4051. Select the A2 by giving the control signal (ABC=010) to MUX 74HC4051 from 8051 uC.
17. Connect +Vcc, GND, and OUT pin of **Precipitation sensor** to +5V, GND, and A3 pin of the MUX 74HC4051. Select the A2 by giving the control signal (ABC=011) to MUX 74HC4051 from 8051 uC.
18. Connect solid state relay input to D3 pin of the NuttyFi to control the hooter.

Figure 18.2 shows the circuit diagram of the system.

18.3 PROGRAM

8051 and NuttyFi are required to program separately for IoT applications.

18.3.1 PROGRAM FOR 8051

```
#include<reg52.h>
#include<stdio.h>
#include<string.h>
#include<stdlib.h>

void delay();
```

```
void transmit(char);
sbit lcd_E  = P2^2;
sbit lcd_RW = P2^1;
sbit lcd_RS = P2^0;

void lcd_init();
void send_command(char command);
void send_data(char databyte);
void delay_lcd();
void display(char, char[]);

char databyte[]={"0123456789"};

char a,b,c;
char x,y,z;
void main(void)
{
TMOD=0x18; // Timer1 Mode2 8 bit auto reload
TH1=0xfd; // 9600 bps
SCON=0x50; // 8 Data bit, 1 start bit, 1 stop bit
TR1=1; // Timer1 ON
while(1)
        {
                unsigned char i;
                i=P1;
                a=i%10;
                b=i/10;
                c=i/100;
                lcd_init();
                P25=0; P26=0; P27=0;
                display(0x80, "Welcome to IoT ");
                display(0xc0, "Wind Sensor:  ");
                send_command(0xc9);        send_data(databyte[c]);
                send_command(0xca);        send_data(databyte[b]);
                send_command(0xcb);        send_data(databyte[a]);
                delay();
                x=P1;

                P25=0; P26=0; P27=1;
                display(0x80, "Welcome to IoT ");
                display(0xc0, "Preciptation:  ");
                send_command(0xc9);        send_data(databyte[c]);
                send_command(0xca);        send_data(databyte[b]);
                send_command(0xcb);        send_data(databyte[a]);
                delay();
                y=P2;

                transmit(x);        transmit(',');        transmit(y);        transmit('\n');
        }
}

void transmit(char y1)
{
        SBUF=y1;
        while(TI==0);
        TI=0;
```

```
}

void display(char x, char databyte[])
{
        char y=0;
        send_command(x);
        while(databyte[y]!='\0')
        {
                send_data(databyte[y]);
                y++;
        }
}

void lcd_init()
{
        send_command(0x03);             delay_lcd();
        send_command(0x03);             delay_lcd();
        send_command(0x03);             delay_lcd();
        send_command(0x03);             delay_lcd();
        send_command(0x03);             delay_lcd();
        send_command(0x03);             delay_lcd();
        send_command(0x28);             delay_lcd();
        send_command(0x28);             delay_lcd();
        send_command(0x0E);             delay_lcd();
        send_command(0x01);             delay_lcd();
        send_command(0x0c);             delay_lcd();
        send_command(0x06);             delay_lcd();
}

void send_command(char command)
{
        char x,y;
        lcd_RS = 0;             /* (lcd_RS) P2.3---->RS , RS=0 for Instruction Write */
        lcd_RW = 0;             /* (lcd_RW) P2.2---->R/W, W=0 for write operation */
        lcd_E  = 1;             /* (lcd_E)  P2.5---->E ,E=1 for Instruction execution */

        x=command;
        x=x >> 4;
        P3=P3 & 0xf0;
        P3=P3 | x;

        lcd_E  = 0;             /* E=0; Disable lcd */
        lcd_RW = 1;             /* W=1;  Disable write signal */
        lcd_RS = 1;             /* RS=1 */

        lcd_RS = 0;             /* (lcd_RS) P2.3---->RS , RS=0 for Instruction Write */
        lcd_RW = 0;             /* (lcd_RW) P2.2---->R/W, W=0 for write operation */
        lcd_E  = 1;             /* (lcd_E)  P2.5---->E ,E=1 for Instruction execution */

        y=command;
        y=y & 0x0f;
        P2=P2 & 0xf0;

        P2=P2 | y;

        lcd_E  = 0;             /* E=0; Disable lcd */
```

```
      lcd_RW = 1;              /* W=1;  Disable write signal */
      lcd_RS = 1;              /* RS=1 */

}

void send_data(char databyte)
{
      char x,y;
      lcd_RS = 1;              /* (lcd_RS) P2.3---->RS , RS=1 for Data Write */
      lcd_RW = 0;              /* (lcd_RW) P2.2---->R/W, W=0 for write operation */
      lcd_E  = 1;              /* (lcd_E) P2.5---->E ,E=1 for Instruction execution */

      x=databyte;
      x=x >> 4;
      P3=P3 & 0xf0;
      P3=P3 | x;

      lcd_E = 0;               /* E=0; Disable lcd */
      lcd_RW = 1;              /* W=1;  Disable write signal */
      lcd_RS  = 0;             /* RS=0 */
      lcd_RS = 1;              /* (lcd_RS) P2.3---->RS , RS=1 for Data Write */
      lcd_RW = 0;              /* (lcd_RW) P2.2---->R/W, W=0 for write operation */
      lcd_E  = 1;              /* (lcd_E) P2.5---->E ,E=1 for Instruction execution */

      y=databyte;
      y=y & 0x0f;
      P2=P2 & 0xf0;
      P2=P2 | y;

      lcd_E = 0;               /* E=0; Disable lcd */
      lcd_RW = 1;              /* W=1;  Disable write signal */
      lcd_RS  = 0;             /* RS=0 */
}

void delay_lcd()
{
      int j;
            for(j=0;j<100;j++)
            {
            }
}

void delay()
{
      int i,j;
      for(i=0;i<=75;i++)
      {
            for(j=0;j<=2500;j++)
            {}
      }
}
```

18.3.2 PROGRAM FOR NUTTYFI

```
#include <ESP8266WiFi.h> // add ESP library
#include "StringSplitter.h" // add string splitter header
```

```
String apiKey1 = "O44YTW0Z5WNO17N8"; // add api key
const char* ssid = "ESPServer"; // add hotspot ID
const char* password = "CCCC@12345"; // add hotspot password
const char* server = "api.thingspeak.com";
WiFiClient client;
String Precipitation,VaneValue,CalDirection,TEMP,PRESS,ALT;
String inputString_Nutty = "";        // a string to hold incoming data

void setup()
    {
    Serial.begin(115180); // start serial at 115180 baud rate
    delay(10); // set delay of 10 mSec
    WiFi.begin(ssid, password); // start Wi-Fi communication
    Serial.println(); // print serial
    Serial.println(); // print serial
    Serial.print("Connecting to ") // print serial
    Serial.println(ssid); //print serial
WiFi.begin(ssid, password); // begin Wi-Fi
while (WiFi.status() != WL_CONNECTED)
    {
    delay(500); // wiat for 500mSec
    Serial.print("."); // print serial
    }
    Serial.println(""); // print serial
    Serial.println("WiFi connected"); // // print serial
    }

void loop()
{
        if (client.connect(server,80))
 {
  NUTTYFi_serialEvent_Nutty();
 send1_TX_NUTTY_PARA();
}
                client.stop(); // stop client
                Serial.println("Waiting"); // print serial
                delay(18000); // wait for 18 Sec

        }

 void send1_TX_NUTTY_PARA() // function to send data to cloud server
 {

      String postStr = apiKey1;
      postStr +="&field1="; // AT command
      postStr += String(Precipitation); // AT command
      postStr +="&field2="; // AT command
      postStr += String(VaneValue); // AT command
      postStr +="&field3="; // AT command
      postStr += String(CalDirection); // AT command
      postStr += String(TEMP); // AT command
      postStr +="&field2="; // AT command
      postStr += String(PRESS); // AT command
      postStr +="&field3="; // AT command
      postStr += String(ALT); // AT command
      postStr += "\r\n\r\n";
```

```
               client.print("POST /update HTTP/1.1\n");
               client.print("Host: api.thingspeak.com\n");
               client.print("Connection: close\n");
               client.print("X-THINGSPEAKAPIKEY: "+apiKey1+"\n");
               client.print("Content-Type: application/x-www-form-urlencoded\n");
               client.print("Content-Length: ");
               client.print(postStr.length());
               client.print("\n\n");
               client.print(postStr);
               Serial.print("Send data to channel-1 "); // print serial
               Serial.print("Content-Length: "); // print serial
               Serial.print(postStr.length()); // print serial
               Serial.print("Field-1: "); // print serial
               Serial.print(Precipitation); // print serial
               Serial.print("Field-2: "); // print serial
               Serial.print(VaneValue); // print serial
               Serial.print("Field-3: "); // print serial
               Serial.print(CalDirection); // print serial
               Serial.print("Field-4: "); // print serial
               Serial.print(TEMP); // print serial
               Serial.print("Field-5: "); // print serial
               Serial.print(PRESS); // print serial
               Serial.print("Field-6: "); // print serial
               Serial.print(ALT); // print serial
               Serial.println(" data send"); // print serial

}

void NUTTYFi_serialEvent_Nutty()  // function to receive serial data
{
    while (Serial.available()>0) // check serial
  {
    inputString_Nutty = Serial.readStringUntil('\n');// Get serial input


    StringSplitter *splitter = new StringSplitter(inputString_NODEMCU, ',', 7);  // new
    StringSplitter(string_to_split, delimiter, limit)
    int itemCount = splitter->getItemCount();

    for(int i = 0; i < itemCount; i++)
    {
      String item = splitter->getItemAtIndex(i);
      Precipitation= splitter->getItemAtIndex(0); // get data of sensor
      VaneValue = splitter->getItemAtIndex(1); // get data of sensor
      CalDirection= splitter->getItemAtIndex(2); // get data of sensor
      TEMP= splitter->getItemAtIndex(3); // get data of sensor
      PRESS = splitter->getItemAtIndex(4); // get data of sensor
      ALT= splitter->getItemAtIndex(5); // get data of sensor
    }
    inputString_NODEMCU = ""; // make string empty
    delay(180); // wait for 180 mSec
  }
}
```

FIGURE 18.3   Window for ThingSpeak.

18.4   CLOUD SERVER

18.4.1   Steps to Create a Channel

1. Sign In to ThingSpeak by creating a new MathWorks account.
2. Click Channels > My Channels, as shown in Figure 18.3.
3. Click New Channel, as shown in Figure 18.4.
4. Check the boxes next to Fields 1–1. Enter the channel setting values as follows:
    Click Save Channel at the bottom of the settings.
5. Check API write key (this key needs to write in the program for local server).

Figures 18.5, 18.6, and 18.7 show the reading of sensors recorded in ThingSpeak server.

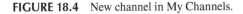

FIGURE 18.4   New channel in My Channels.

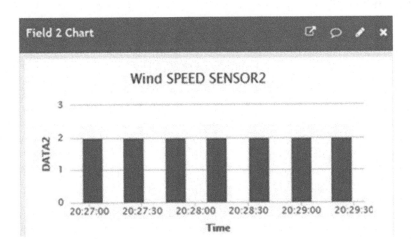

FIGURE 18.5   Wind speed.

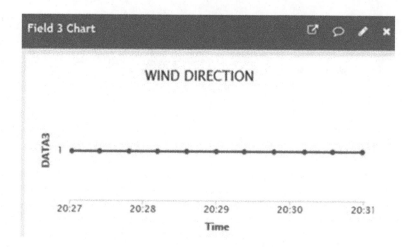

FIGURE 18.6   Wind direction.

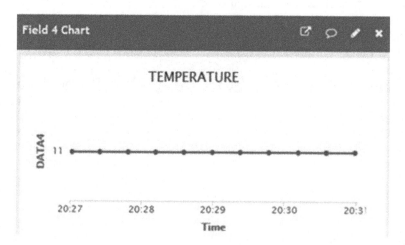

FIGURE 18.7   Temperature.

Index

Page numbers in *italics* refer to figures, and those in **bold** refer to tables.